GREEN UP!

Green Up!

An A-Z of environmentally friendly home improvements

WILL ANDERSON

green books

Printed by T J International, Padstow, Cornwall, UK
on 100% recycled paper (100% post-consumer waste)

DISCLAIMER: The advice in this book is believed to be correct
at the time of printing, but the authors and publishers accept
no liability for actions inspired by this book.

ISBN 978 1 903998 94 6

Contents

Introduction

There's no place like home to start greening up your life:

- We burn more energy in our homes than in industry, on the roads, or flying.
- We each produce over half a tonne of waste in our homes every year, only a quarter of which is recycled.
- Every day each of us despatches over 150 litres of domestic drinking water down the drain.

We use more energy, water and resources in our homes than we have ever done, but the modern world is so tidy and well organised that it is easy not to notice. Unfortunately the environmental impacts of our daily consumption are just as hidden, so it is quite possible to potter through each day and absent-mindedly wreck the planet. Thankfully, however, it is possible to reduce these environmental impacts without sacrificing the pleasures of domestic life. Just about everything you can do to green up your home will also benefit you: lower utility bills and a warm, comfortable, healthy home are the immediate rewards.

There are lots of small steps you can take without spending any money at all, such as turning the thermostat down a degree, turning off lights and appliances when they are not needed, turning off the tap when brushing your teeth and recycling your kitchen waste. But if you want to make a bigger difference, you need to set about improving the environmental performance of the space you live in.

Happily, home improvements are all the rage these days, so whenever you have a job to do, you can choose a greener solution.

Every domestic DIY job is an opportunity to green up: buying a fridge, painting the hall, changing a light bulb, fixing the shower.

But don't stop there. Consider all the ways you can reduce the energy, water and other resources that your home consumes. Some options are challenging and expensive (and may require planning permission), whereas others are remarkably simple and cheap. Although renewable energy technologies such as wind turbines and solar panels grab the headlines, it's always best to reduce the energy you consume before you start making your own. Insulation, draught-proofing and low-energy light bulbs do a great job and will save you serious amounts of money.

Once you have done what you can to save energy and water, you can start thinking seriously about providing for yourself using the abundant resources of nature: the sun, wind and rain. Renewable energy technologies can seem a bit mind-boggling at first, but this book will help you understand both how they work and what they require to operate effectively. Not every technology is suitable for every household, so it's always important to carefully consider the best options for your circumstances.

There are various national and local grants available for energy efficiency improvements and renewable energy systems. As these grants have been very prone to change, this book does not provide detailed information about what is available. Fortunately the Energy Saving Trust and local energy advisors continue to provide excellent free advice about them.

Don't be put off by the sheer range of things you can do. Every step is important, however small. Begin by greening up the way you think about your home and how you live in it. In time your home will follow suit, becoming a better place both for you and for the world you live in.

Boilers

If you have a cranky old boiler in your back cupboard, replace it. A modern boiler will cut your energy bills and significantly reduce your carbon emissions. Four-fifths of the energy we burn in our homes goes on heating and hot water, so a high performance boiler is a must-have for an environmentally friendly home.

Choosing a boiler

Boilers heat up water for your taps and central heating by burning fuel. The fuel you burn is your first consideration when choosing a heating system. If you are on the gas main, you are already plumbed into the cleanest and most convenient fossil fuel. If you rely on deliveries of oil or LPG (liquefied petroleum gas), you might want to consider burning wood (page 132) or installing a heat pump (page 51), as they produce even lower carbon emissions than natural gas. If you heat your home with electricity, you are using the dirtiest of all fuels because so much energy is wasted in the power station (page 32). Switch fuel if you can, or use your electricity to run a heat pump.

Whatever fuel you use, buy the most efficient boiler to burn it. There are hundreds of boilers on the market, but you can narrow your options by using the Boiler Efficiency Database at www.sedbuk.com (SEDBUK stands for Seasonal Efficiency of Domestic Boilers in the UK – a standard against which all boilers are rated). Go for an A-rated condensing boiler which will have an efficiency of at least 90% (many old boilers are only 60% efficient). Alternatively, look out for the blue 'Energy Saving Recommended' label, which is only given to the most efficient boilers.

Condensing boilers

Condensing boilers used to be treated with suspicion by plumbers, but happily they are now a legal requirement so you should have no difficulty finding an installer. They extract so much heat that the combustion gases sometimes condense. Although in most situations condensing boilers can directly replace an old boiler, an extra bit of plumbing is required to drain the liquid condensate. Condensing boilers work well with both radiators and underfloor heating.

Hot water cylinders and combination boilers

Most boilers provide hot water via a hot water cylinder. This allows the boiler to run at capacity for a sustained period, heating up a large amount of water which is then drawn as needed from the cylinder. The only problem with this design is the heat loss from the cylinder. Although this heat can be useful, for example in an airing cupboard, much of it is wasted, especially in summer. To reduce this heat loss, make sure that your cylinder has one or more well-fitting cylinder jackets and the pipes coming off the cylinder are insulated for at least a metre. It is best to heat up the water in the cylinder at the times of day when you need a lot of hot water, using a timer (see page 12), rather than keeping it hot all the time.

Combination boilers (combis) heat water on demand rather than storing it in a hot water cylinder. They are more energy-efficient because the background cylinder heat losses are removed. However, avoid combis that come with a 'keep hot function' as they waste energy in order to deliver ultra-quick hot water.

There is one big green reason for having a hot water cylinder: you can heat the water in it with a solar hot water panel as well as your boiler. If you are upgrading your heating system, now is the time to think about installing a solar panel and getting free heat from the sun (page 91).

Sizing a boiler

If you are making improvements to the insulation and airtightness of your home, you may dramatically reduce your demand for space heating. Consequently you may not need such a big boiler. A boiler that is too big for its boots will not operate efficiently, so try to avoid over-sizing your new boiler. There is a guide to sizing your boiler on the SEDBUK website.

Costs

As condensing boilers are now required by law, prices are competitive. A-rated boilers start at around £700. Insulated cylinder jackets start at around £15.

Further information

Talk to an energy efficiency advisor before you upgrade your boiler or heating system. Call the **Energy Saving Trust**'s helpline on 0800 512 012 to be put in touch with a local advisor. The Trust's website offers useful information on boilers and heating systems (www.energysavingtrust.org.uk) and its free online library of best practice guides includes *Domestic Condensing Boilers: The Benefits and the Myths* and a series of technical guides on domestic heating systems for gas, oil and solid fuel.

Central heating

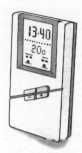

Domestic life in the bleak midwinter used to be centred on the main room of the home, heated by an open fire. Family members would retreat to their bedrooms only when necessary, diving under heavy blankets when they got there. Bathrooms, if they existed, were not places of relaxation. For most of us, central heating has swept all this away, but at a price: higher energy consumption and a whole lot more carbon emissions. You can, however, cut costs and emissions by operating your central heating as efficiently as you possibly can.

The right temperatures

At the heart of any central heating system lies a heat source: your boiler. If you have an old boiler, your first thought should be to replace it with a more efficient model (page 8). Then make sure that your central heating is taking only as much heat from the boiler as you need to maintain the temperatures that you want in your home. An increase in room temperature of just 1°C can cause a 6–10% increase in fuel costs, so good controls can make a real difference to your bills.

The most important heating control is your room thermostat: the wall-mounted dial or digital panel that sets the primary temperature for your home. This thermostat controls the boiler and keeps it running until the temperature you have set is reached. It should be positioned away from heat sources such as radiators, fires and direct sunlight.

If your boiler keeps on firing up and staying hot even when no central heating or hot water is needed, get an electrician to rewire the thermostat to prevent this happening as it is very wasteful. This wiring is called a 'boiler interlock'.

In the right places

The other vital temperature controls for central heating are thermostatic radiator valves (TRVs). If you fit these to your radiators in place of the basic on/off flow valves, you can set specific temperatures for each radiator and so for each room. This means that you do not need to heat your whole house to the temperature set on your main room thermostat. They are also helpful in rooms which get heat from other sources, particularly direct sunlight, as they prevent the room overheating by shutting off the radiator as soon as the desired room temperature is reached. Do not fit TRVs in the room where your main room thermostat is located as the thermostat will control the temperature of this room directly.

At the right times

It is often said that leaving the central heating on all the time is cheaper than switching it on and off again. This is not the case. A programmer allows you to set the times when the heating automatically turns on and off, matching the times of day when you actually need to heat your home. Sometimes this is combined with the room thermostat in a programmable room thermostat.

A good programmer will be clear and easy to use. It will allow you to set different heating times for your central heating and hot water and different daily heating schedules for weekdays and weekends. Simple boost functions allow you to extend the heating period on a one-off basis without having to re-programme the timer.

Central heating thermostats and programmers are often poorly designed and hard to use. As a result many people ignore them and use their living room fires to provide supplementary heating and open windows to cool down. As this is very wasteful, learning to use your programmer properly is one of the quickest ways of saving energy and money in the home.

Upgrading your controls

Installing a new programmable room thermostat is a relatively simple wiring job. However you need to be sure that your new controls are compatible with your existing wiring. Get technical advice from the companies selling the equipment or talk to an electrician first. Installing TRVs requires at least partial drainage of the central heating, so help from a plumber is recommended.

If your central heating is old, get professional advice to see how the performance of the whole system can be improved.

Getting the most from your radiators

A simple way to improve the performance of your radiators is to put insulated foil panels behind them to prevent them heating your walls (and whatever lies beyond them) rather than the room. You should also ensure that radiators sited below windows do not get covered by curtains at night as this will direct the heat out of the windows rather than into the room.

If you are doing a major renovation, consider installing underfloor heating as this is more comfortable to live with and more energy-efficient (page 104).

The fireside

In many centrally heated households the fire in the main living room is still used. As most fires are inefficient when compared to modern boilers, it's best to avoid this if you can and try to set temperatures throughout your house using your central heating controls. Coal and electric fires are the worst, and gas 'coal effect' fires should also be avoided. A fire that you turn on just for something to look at is a particularly poor use of energy.

A wood stove, however, is very environmentally friendly (page 136). If you have a wood stove or wood fire, try to keep it burning and your central heating turned off for as long as possible.

Costs

Programmable room thermostats start at around £40. Look for a product with the blue 'Energy Saving Recommended' label.

Thermostatic radiator valves cost £10–£20. Insulated foil panels for radiators cost less than £10.

Further information

Talk to an energy efficiency advisor before you upgrade your heating system. Call the **Energy Saving Trust**'s national helpline on 0800 512 012 to be put in touch with a local advisor. The Trust's website also offers useful information on heating systems and their controls (www.energysavingtrust.org.uk) and its free online library of best practice guides includes the *Domestic energy primer – an introduction to energy efficiency in existing homes.*

Conservatories

One of the easiest ways to increase the size of your living space is to put a conservatory on the side of your house. But conservatories were originally designed for plants, not people, and we humans are more sensitive to temperature swings than the tomatoes we tend. Consequently life in a conservatory can be an unpleasant experience unless lots of energy is burnt to maintain comfortable conditions – a financial and environmental disaster.

Avoid the quick fix

Conservatories are poor living room extensions because they overheat in summer and stay chilly in winter. If you want more living space, avoid the quick fix by building an extension that will be comfortable and enjoyable throughout the year without requiring lots of heating and air-conditioning. You may still want to include significant glazing to bring more light into your home, but this can be balanced with a well insulated roof and walls.

Go with the sun

As conservatories heat up quickly in the sun, they are best used as daytime sitting rooms which you retreat from when the sun goes down, for they cool just as rapidly. If you use a conservatory in this way you do not need to heat it. The pleasure of the conservatory becomes its seasonality, your use of it following the daily and annual patterns of free solar heat.

An unheated conservatory is the only green choice. However, be sure that the addition of an unheated conservatory does not lead to increased heat losses from the room it adjoins. The wall between the two rooms must be fully insulated, any partition glazing must be the best you can afford, and the adjoining door should be as airtight as any other exterior door.

Reduce active heating

If you already have a heated conservatory and do not want to turn it into an unheated space, do what you can to reduce the amount of heat it wastes in winter. Install thermostatic valves on the radiators to ensure that they shut down when the conservatory catches the sun. These can also be used to turn the radiators off when the conservatory is not being used. Draught-strip the doors and windows (including the internal door) and close the blinds at night to help keep the heat in.

If you are renovating or building a new conservatory and are determined to use it throughout the year, specify triple glazing and airtight doors and windows. Big conservatories of 30m² or more are now covered by the energy standards in the building regulations.

Reduce active cooling

As our summers get hotter, so our conservatories become ever more tropical and the temptation increases to install or hire air-conditioning. Avoid this with careful use of shading, ventilation and materials.

Most conservatories come with shading on the inside. This reduces glare but lets most of the heat in. Light-coloured blinds reflect more of the light back out before it turns to heat, but the best option is shading on the outside. This is harder to install and control than interior blinds, but fixed shading can be used selectively to reduce the heat gains without major loss of light. A timber lattice and/or deciduous climbers on the roof are particularly effective.

Make sure you can open windows on opposite sides of the conservatory or in its roof to flush out excess heat. A floor made of stone flags, dark tiles or bricks will help absorb the heat of the sun and reduce the swings in temperature (see also Cooling, page 21).

Reduce your heating bills

Many ultra-low-energy houses are built with unheated 'sunspaces' which help to warm the rest of the house, especially in the spring and autumn when the need for heat is matched by plentiful sunshine. You can make an unheated conservatory work for you in this way by letting the warm air that builds up in the conservatory flow into the rest of your home, then shutting off this flow as soon as the air temperature in the conservatory drops.

An unheated conservatory with good thermal protection from the rest of the house also acts as a buffer zone in the winter, slowing the penetration of cold into the building.

Further information

A free information sheet on conservatories is available from the **Centre for Alternative Technology** (www.cat.org.uk, 01654 705989). The **Energy Saving Trust**'s free online library of best practice guides includes *Energy Efficient Domestic Extensions* (www.energysavingtrust.org.uk).

For more information on trapping the heat of the sun to help heat your home – 'passive solar design' – see the *Green Building Bible* (Green Building Press, www.greenbuildingbible.co.uk) or *Ecohouse* (Sue Roaf, Architectural Press).

Cookers

Cooking is one of the more obvious ways that energy is burnt in our homes. Every hot dinner you consume generates carbon emissions from the cooking fuel as well as aromatic emissions from the food. So although cooking may not be the biggest energy burner in the home (that's your heating), it's an easy one to green up if you are fitting a new kitchen. Grab the opportunity to fit energy-efficient technology, then go greener still with energy-aware cooking and shopping.

Put the heat in the pie

If you want to cook an apple pie without wasting energy, you should do your very best to heat the pie – not the dish, the oven and the kitchen. Think of energy efficiency in this way, and the heroes and villains of cooking technology are easy to spot.

A microwave oven heats the pie without the oven heating up first. Nothing is wasted in the transfer of energy from the oven to the food, and there is no time lag for the heat to get to the centre of the dish. Not surprisingly microwaves are the most energy-efficient ovens you can buy (though the crust of the pie may not be to your liking).

The best traditional ovens stop heat leaking into your kitchen with excellent insulation (look for thick oven walls) and a double-glazed door that closes tightly. In contrast, an oven that is left on all the time is an energy disaster: a range oven that quietly burns energy throughout the summer is senseless.

Most hobs work by a) getting hot and b) transferring the heat to the pan, which then transfers the heat to your food. The exception is the induction hob which works by creating heat in the pan through electromagnetic induction. As the hob itself does not get hot, except through contact with the hot pan, this is the most efficient technology available, though it can only be used with iron or steel pans (anything that a magnet will stick to).

Fuel choice

Your choice of fuel makes a big difference to your culinary carbon emissions. Wood-fired ovens are the greenest choice , but are rarely practical. For most of us, the best option is to cook with natural gas, LPG (liquefied petroleum gas) or oil as, unlike electricity, the fuel is burnt at the point of use. Electricity is always the dirtiest fuel because so much energy is lost in the power station (see page 32).

The 'carbon intensity' of electricity takes some of the green sheen off microwaves and induction hobs. However, even on a carbon comparison, microwaves come out top and induction hobs are similar to gas. So the greenest kitchen would have a microwave, a well-insulated gas oven and either a gas hob or an induction hob.

All electric ovens come with an energy label which tells you how energy-efficient the oven is on an A to G scale. This makes it easy to choose a very efficient (A-rated) electric oven. Unfortunately gas ovens do not have energy labels, so don't be put off by the apparent lack of A-rated gas ovens. The high carbon intensity of electricity means that if you have a gas hob, it is better to use this to heat water rather than relying on an electric kettle.

Green cooking

You can also save lots of energy by cooking with care. Keep lids on pans and use a pan that covers the entire ring or flame. Steam or pressure cook where possible and thaw foods completely before cooking, preferably in a fridge as this helps to keep the fridge cool. Turn the oven or hob off before the meal is cooked: to boil an egg, put the egg in cold water in the pan, bring to boil, turn off and leave for five minutes with the lid on.

Food energy

The amount of energy required to manufacture and deliver food is many times greater than the energy you get from eating it. So if you want to green up your cooking as well as your cooker, you should do everything you can to buy locally produced food that is minimally processed and packaged. If you can grow your own food, you can cut out these huge energy costs altogether.

Costs

Higher-quality appliances cost more, but perform better and last longer. So try to think about cooker costs in the round. A-rated electric ovens are available in all price bands. The range of gas ovens is more limited, but any gas oven will be much cheaper to run than an electric oven. Induction hobs are expensive, starting at about £500 compared to £100 for the cheapest electric hobs, but they are exceptionally quick, energy-efficient and economical.

Further information

For ideas and tips on low energy cooking, see www.theyellowhouse.org.uk. The **Soil Association** provides an online directory of organic shops and local suppliers including box schemes and farmers markets (www.whyorganic.org).

Cooling

In chilly Britain we use lots of energy to keep us warm in winter and very little to keep us cool in summer. But this is changing: hotter summers, greater wealth and a growing supply of domestic air-conditioning units are pushing up our summer electricity use and making the planet hotter still. Happily, however, you can stay both cool and green by prioritising simple, passive ways of managing heat in your home.

Keep the heat out

The first line of defence is the best: stop the heat getting in. In southern Europe most houses have external shutters for this purpose but you can also use external blinds, fixed window overhangs, awnings, horizontal louvres or even deciduous plants that provide shade in the summer and cast off their leaves in the autumn when you want the sun's warmth back.

Although south-facing rooms suffer the most from overheating, they are relatively easy to shade because the high summer sun does not penetrate very far into the room. A west-facing room is trickier because even at midsummer the sunlight will get deep into the room. Horizontal window overhangs or louvres are not very effective here, so use blinds, shutters, vertical trellises or carefully placed trees instead.

Curtains and internal blinds do not stop sunlight entering a room, where it turns to heat. Nonetheless, light-coloured blinds reflect some of the light back out again. Closing your curtains on hot summer days is a good way to keep cool.

Insulation not only keeps heat in during the winter months, it also keeps heat out during the summer. The most uncomfortable rooms during the summer are often upstairs bedrooms, underneath poorly insulated roofs. If you improve the insulation in your roof (page 76) you will enjoy cooler nights as well as lower heating bills.

Energy-efficient lights and appliances also help to keep the temperature down because they produce very little heat.

Keep the air moving

Our first response to an overly hot room is usually to open a window and bring in some fresh air. However this is only good idea if the air inside your home is warmer than the air outside. The best time to cool your home this way is in the evening or at night when the outside air is cool.

Hot air rises, so if you open a window on the ground floor and a window on an upper floor and leave the internal doors open, your house will gently flush the warm air out. You can also encourage cross-ventilation in single rooms or on single floors by opening windows at opposite ends of the room or floor. Windows that can be left open securely overnight, even by only a crack, are the best means of summertime cooling.

Natural ventilation through your house also creates a movement of air across your skin which is pleasant and cooling. Electric fans work in the same way, at a higher speed, but do not replenish the air. Try to stick to the natural method if you can.

Dampen temperature swings

Most of us enjoy the sunshine pouring in to our living spaces, and don't want to block it out altogether. You can reduce the risk of overheating in such rooms by using exposed heavyweight materials that absorb the heat during the day and release it over the evening and night. Masonry walls do this job, as do concrete or stone floors. If you have a heavyweight floor, avoid rugs or carpets as they impede the heat being taken up by the floor.

Cool the air

Water is used in hot countries to keep buildings cool because it absorbs a lot of energy when it evaporates. If you don't have space for an internal water feature, use houseplants instead as they do the same job through transpiration.

Costs

External shading of all kinds – shutters, blinds, fixed louvres and pergolas – must be well made in order to cope with long-term abuse from the sun, wind and rain. These can be costly choices, especially when made-to-measure. The more complex the solution, the higher the cost: motorised external blinds are expensive whereas a simple fixed overhang is relatively cheap. A pergola for your French doors will cost a minimum of £100.

It costs nothing to open your windows, though you may need to upgrade your handles and locks if you want to leave a ground floor window open overnight. Stone and tile floors are more expensive than carpets, but far more durable.

Further information

A free sheet, *Keeping Cool in Summer*, is available from the **Centre for Alternative Technology** (www.cat.org.uk, 01654 705989).

Dishwashers

Dishwashers are no longer environmental outcasts. Improvements in their energy and water efficiency mean that running a dishwasher can now be more environmentally friendly than washing by hand. However, this depends: there are many ways to wash the dishes, and many ways to use a dishwasher. If you are investing in a new machine, you need to match the best technology with a careful approach to using it.

The art of washing up

When you hand-wash your dishes, how do you go about it? Do you use lots of running hot water, rinsing as you go and changing the water in the sink regularly? Or can you clear an entire dinner party with one bowl of hot water? If you follow the latter method, switching to a dishwasher will not lead to a reduction in your water and energy use. If you are less meticulous, a high quality dishwasher may be a greener choice.

The very best full-sized dishwasher uses only ten litres of water, less than is needed to fill a typical kitchen sink. Its energy consumption is around one kilowatt-hour per load. You will need a big sink of hot water to match this but change the water a couple of times and the dishwasher has got the better of you. Don't forget, though, that the manufacture of the dishwasher also consumes a lot of water, energy and resources. This is why durability matters just as much as energy and water efficiency when it comes to choosing a dishwasher (see Washing machines, page 116).

Choosing a dishwasher

All dishwashers come with a triple letter rating on the energy label, such as 'AAA'. The first letter is the most important as it describes energy efficiency. The second letter describes washing performance (how clean the dishes come out) and the third letter describes drying performance (how dry the dishes come out). The energy label will also tell you how much water is used in a standard cycle, another important point of comparison.

As many dishwashers now have an A energy rating, look more carefully at the energy label to see how much energy each machine actually uses on a standard cycle (e.g. 1.05 kWh). You can then ensure that you are choosing the most energy-efficient of the A-rated models. The blue 'Energy Saving Recommended' label is only given to the most energy-efficient dishwashers, so look out for this too.

If you can afford it, go for a high quality brand that will perform well and last for many years. Make sure that the machine offers a good choice of cycles including an energy-saving option.

Using a dishwasher

The following tips will ensure that you get the very best environmental performance from your dishwasher:

- Never rinse plates first. This is unnecessary for modern machines and very wasteful. Just scrape off the worst food remains.
- Keep your salt reservoir topped up, especially in hard water areas, to ensure that the heating element and the jet sprays work efficiently.

- Always wash a full load.
- Omit or stop the drying stage of the cycle and leave the door ajar for natural drying.
- Avoid harsh detergents. Use an eco-friendly detergent designed for a dishwasher (not a washing-up liquid).

Costs

A top of the range dishwasher costs between £500 and £1,000. The more expensive brands are also more expensive to repair, but they should break down far less often than the cheap models and last much longer.

Further information

A detailed ranking of dishwashers based on water and energy efficiency is provided by **Waterwise** (www.waterwise.org.uk).

A guide to using and maintaining dishwashers can be found at www.dishwasher-care.org.uk.

Draught-proofing

Nothing spoils a cosy room quite like a draught. Even if the air temperature in your home is high, you will still feel a chill if there are currents of cold air ebbing and flowing around you. As the cold air rushes in the warm air rushes out, so draughts are responsible for significant heat losses from your home. Happily, many draughts are very easy to prevent. If you have never done any draught-proofing, this is an effective and economic place to start greening up your home.

Uncontrolled draughts and controlled ventilation

On average, one-fifth of the heat lost from a home in winter is due to air escaping though all the holes in the building. Stopping up the holes is a relatively easy task, with immediate benefits: draught-proofing will save you money and make your rooms warmer, more comfortable, quieter and less dusty.

If you are very thorough in your draught-proofing, you may need to improve your ventilation (page 107). It might seem odd to stop up the holes in the building only to open new ones elsewhere but your aim should be to reduce the air movement that is out of your control and replace it with ventilation that you can control. By doing this you can get fresh air where you need it and when you need it without throwing heat away unnecessarily.

In practice, even if you plug all the obvious gaps a lot of fresh air may still get in along invisible pathways such as minute holes in brick and concrete walls. Consequently you may be able to radically reduce your draughts without having to install extra ventilation.

Never block up air-bricks or other vents in rooms with fires or old boilers which draw oxygen to burn from the air in the room. Modern boilers draw fresh air from outside the building through dedicated ducts, so if you upgrade your boiler you can also improve your draught-proofing.

Materials

Draught excluders include self-adhesive foam or plastic strips, rubber and brush seals in aluminium or plastic carriers, compression strips which tighten when a door or window is shut against them, and piped liquid sealants applied with a sealant gun. Have a look at the materials available in your local DIY store and work out which are most suitable for your various draught-proofing jobs.

Exterior doors

There are various ways of closing the gap at the bottom of a door, including a brush strip fixed to the door, a compression or rubber seal attached to the sill, or an interlocking system with separate pieces for both door and sill. The gap around the rest of the door can be sealed with foam, rubber or brush strip. Air will also get through the junction of the door frame and the wall, so close this with a piped sealant. And don't forget a keyhole cover and brush excluder for the letterbox.

Windows

The most effective way of cutting the draught through an old window is to install secondary glazing (page 130). Alternatively, do your best to draught-strip the gaps where the window closes against the frame with a foam or rubber strip attached to the frame. For sash windows, fit sprung or brush strips to the frame where the window slides against it and rubber strips at the top and bottom. You may also need a brush strip at the top of the lower sash where it meets the upper sash. If you can afford it, a better option is to get a professional joiner to renovate the sashes and install recessed draught excluders as part of the job. If there is any sign of cracks between your walls and the window frames, close them with sealant.

Floors and ceilings

If you have timber floorboards laid over timber joists, lots of air can pour through the floor either from under the house (on the ground floor) or via the ends of the floor cavity (on upper floors). The best solution is to take the floorboards up and pack insulation between the joists. This is essential on the ground floor as you need insulation to prevent heat loss to the ground. A cheaper option on upper floors is to fix a new hardboard layer over the floorboards with a final finish on top. If you want to keep the exposed floorboards, close the gaps with a sealant or use a product designed to be pushed into the gaps. Take your sealant gun to the top and bottom edges of the skirting boards and do your best to seal light fittings or pipes that penetrate ceilings.

Service pipes, ducts and holes

Examine all the points at which services come into or exit the building. Chimneys are the biggest holes: if you do not need them to vent gas or open fires, block them up. A partially inflated balloon is a simple temporary way of doing this. Use sealant to close gaps around water and waste pipes, boiler flues and extractor fans. For larger gaps, bigger than 6mm, attach foam strip first to support the sealant or use expanding foam. Do the same with gas and electricity ducts and cables. Use sealant on cracks around light switches and power points.

If you are doing any major works, such as installing a new kitchen or bathroom, take the opportunity to check all these possible leaks, especially the ones that will be out of sight when the room is complete. A plaster finish is a good way of providing a final seal for all your walls.

Loft and basement

If you do not use your loft or basement as regular rooms, they should be kept outside the protected warmth of the building. If so, treat loft hatches and basement doors as external doors and draught-strip them carefully. Bolts may be needed to keep them securely closed. Insulation packed between the joists in the loft helps to keep draughts out as well as heat in (page 56).

A draught lobby

In the winter you can lose a lot of heat through your front door if you stand with it open when you are taking your boots off, chatting to the postman, saying farewell to your friends or dismantling your pram. To avoid this, build a porch with inner and outer doors (a draught lobby). You can then pursue all these activities quite easily without ever opening both sets of doors at once. Alternatively, line up your guests at the front door, bid them farewell, then shoo them out.

Pressure testing

For the ultimate in air-leak detection, get your home pressure-tested. This procedure involves closing all your windows and doors and blowing air into the building with a very large fan. Using a little smoke puffer you can then identify precisely where all the air leaks are. It is an expensive procedure but worth considering if you are doing a major renovation and want to radically improve the performance of the building.

A cheap alternative is to wait for a windy day, then go round your house with a smoking incense stick and watch what happens to the smoke.

Costs

Many of the tasks of draught-proofing can be achieved at low cost. Typical prices include: brush door strip, £5; threshold rubber compression seal, £15; brush letterbox excluder, £4; and self-adhesive rubber strip, 50p per metre.

Further information

Contact the **National Insulation Association** (01525 383313, www.nationalinsulationassociation.org.uk) to find an insulation and draught-proofing installer.

The **Energy Saving Trust**'s free online library of best practice guides includes *Energy efficient refurbishment of existing housing*, with a chapter on draught-stripping and guidance on materials (www.energysavingtrust.org.uk).

Electricity

Electricity is fantastically convenient and clean at the point of use, yet it is the dirtiest of all the fuels we use in our homes because so much energy is wasted in the power station. Out of sight and out of mind (for most of us), power stations relentlessly spew carbon dioxide and toxic particles into the air. Greening up your electricity use is therefore a top domestic priority.

National electricity generation

Most of the electricity produced in Britain is generated in coal (37%), gas (35%) and nuclear (19%) power stations. Only 4% comes from 'renewable' sources, and much of this is from burning the gas that comes off landfill waste dumps. The supply of truly sustainable electricity, such as wood, wind and solar power, is increasing but is still very scarce.

Coal and gas power stations are the single biggest source of carbon dioxide emissions in the UK. Modern gas-fired power stations are the most efficient of the fossil-fuel stations, but half the energy in the gas is still lost up the power station chimney as heat rather than converted to electricity. Nuclear power stations have low carbon emissions but create an ever-growing stockpile of radioactive waste that will remain toxic for thousands of years (long after all current civilisations have disappeared). Whatever the pros and cons may be of fossil fuels and nuclear power, neither can be considered green.

Greener electricity

There are three ways you can reduce the environmental impact of your electricity consumption: use less, make your own and switch to a renewable supplier. The first of these is by far the most important. As we will never be able to meet our current national demand for electricity using renewable power alone, a truly green future for electricity is only possible if we radically reduce demand as well as increasing renewable supply. If you can also boost this supply by installing some renewable power of your own, so much the better.

Reduce demand

Everyone can take steps to reduce their use of electricity. This book is full of suggestions for cutting the energy needed at home for lights, appliances and electronic equipment. Whenever you replace an electrical appliance, buy a model with the very best energy performance (remember that the A-G scale on the energy label for some appliances has been extended to A+ and A++). There are also lots of simple behavioural changes that will reduce your electricity use at home, such as switching off lights when you leave a room, turning off televisions and other electronic equipment at the plug, and only filling the kettle with the amount of water you require.

If you use electricity for heating and hot water, consider whether you could switch to a different fuel. It is far more efficient to burn gas in a boiler to heat your home than to burn it in a distant power station and then use the electricity generated for the same purpose (see Boilers, page 8).

When trying to reduce your electricity use, it helps if you can see the difference your actions make. Unfortunately many electricity meters are hard to read and hidden in cupboards, but you can buy user-friendly and hand-held meters which tell you how much electricity you are using and so help you identify where savings can be made.

Make your own

Most households in Britain could generate some or all of their own electricity. If you have a sunny roof or wall, solar power is within your reach (page 91). If you have an unobstructed view of the prevailing wind, a wind turbine is definitely worth considering (page 124). If a river runs through your back garden, a water turbine is a possibility (page 120).

Domestic micro-power stations can be fully integrated with the national grid. This means that when you are generating more power than you need, the surplus electricity is exported. When you need more electricity than you are generating, you buy it back. This is a completely automatic and seamless process. It is possible to store surplus power in batteries, but this is not recommended if you are grid-connected. Batteries are not eco-friendly and they also take up space, so it is better to use the grid as your battery and be paid for any surplus you make.

Energy companies offer a variety of deals for households who generate their own electricity, so shop around. Ideally the price you are paid for the electricity you export should be the same as the price as you pay for the electricity you import. However, some companies offer a lower price that covers everything you generate, not just what you export.

Some power companies will also register you to receive Renewable Obligation Certificates (ROCs) from the government and pay you for the ROCs you are awarded. In Britain every power company has a legal obligation to produce some of its power using renewable sources of energy. However power companies that do not generate enough can buy ROCs from companies that produce more than their legal obligation. Although designed for big power companies, households that generate renewable power can also receive ROCs. The value of ROCs is determined by the market, but you could make upwards of £100 per year.

Green electricity suppliers

Switching to a green electricity supplier or on to a green tariff is relatively easy to do, as you do not need to make any physical changes to your meter or wires. But this doesn't mean that you can then forget about your electricity consumption because it is all 'eco-friendly'. Renewable electricity is a scarce and precious resource. If you waste it, you are preventing others from using it and forcing them to use electricity made in coal, gas and nuclear power stations. As long as renewable electricity remains scarce (the foreseeable future), we must all use it with as much care as possible.

In practice, many green tariffs do not provide 100% green electricity. Some promise to pay money into green projects, and others will plant trees to 'offset' the carbon emissions of your electricity consumption. Although no provider can guarantee that the actual electricity you use comes from renewable generation, ideally a green provider will put as much renewable power into the national grid as you draw from it.

Research your options before you choose a green tariff, and make sure you know exactly what the supplier is promising to do with your money. But don't be put off by the confusing range of choices on offer. For all the problems of green tariffs, they are still worth signing up for.

Costs

User-friendly electricity meters include the Electrisave (www.electrisave.co.uk), Efergy meter (www.homeco2meter.com) and the Wattson (www.diykyoto.com). Expect to pay £45 or more.

Some green tariffs cost no more than ordinary electricity tariffs, but green tariffs with premiums tend to provide greater environmental benefits. If you are willing to pay more, make sure your money is being put to good use, for example in building new renewable electricity generating capacity.

Further information

The **Energy Saving Trust** maintains a database of Energy Saving Recommended products (www.energysavingtrust.org.uk, 0800 512 012). In shops, the products included in this database will display the blue Energy Saving Recommended label as well as the standard energy label. For the very best models on the market see www.topten.ch.

Energywatch (www.energywatch.org.uk) maintains a list of green tariffs with details of what each offers.

Companies which provide comparative information on green electricity tariffs and assist in switching include www.uswitch.com, www.greenelectricity.org and www.energylinx.co.uk.

Floors & floor coverings

Of all the things that make a home, none work quite as hard as our floors. We abuse them thoroughly every day, yet expect them to play a leading role in defining the look and feel of our living spaces. If you are making changes to your floors, be sure that they can cope with this abuse and stay looking good for many years. And don't forget the equally important role that your floors should play in keeping you warm in winter.

Stop the heat loss first

Before you start putting a new floor or carpet down on your ground floor, consider what lies below. If you have a suspended timber floor – floorboards supported by wooden joists sitting above a ventilated void – this is the ideal time to radically reduce your ground floor heat losses by fitting insulation beneath it.

This is not as difficult as it might seem, though it does mean taking up the floorboards. Netting is then fixed to the bottom of the joists to support the insulation which is packed in, filling the gaps between the joists. This approach maintains a space below the insulation, keeping the air flowing and preventing the build-up of damp (make sure you do not block any air bricks beneath the floor). Air that passes under the insulation, rather than below and through your floorboards, does not carry away the heat from the room above. You will appreciate the warmth and lack of draughts straight away.

Draughts can also be a problem in upper floors, caused by air flow through the ends of the floor cavity. To prevent this, use sealant to close gaps in any exposed floorboards and in the junction of the skirting board with the wall and the floor (page 29).

Durability is all

Cheap floors and carpets show their age all too soon and are regularly ripped out and replaced. Consequently large amounts of energy and resources are consumed making new floor coverings while the old ones pile up in landfill sites. Whatever you want to feel underfoot, do what you can to ensure that the feeling will last for as many years as possible.

Hard floors last the longest

The best eco-options are hard floors made of locally sourced natural or reclaimed materials. Fortunately, materials such as stone and wood tend to look good and improve with age, so these are great choices if you can afford them. Ceramic or terracotta tiles also look great and last forever, but more energy is needed to make them because they are fired.

Much of the stone flooring on the market today is imported from India, China and Brazil. As stone is so heavy, these imports leave a significant trail of transport emissions behind them. Wherever possible, choose British stone.

Timber floors are likely to come from Europe unless you are seeking a tropical hardwood. It is always a good idea to ensure that the timber you buy comes from a certified sustainable source (page 132) but this is essential if the wood comes from the tropics. Timber floors should always be given time to adjust to the humidity of a house before they are laid.

Reclaimed materials are the very best eco-option because you are keeping old resources in use rather than extracting new ones. There is a good supply of reclaimed timber, stone and tile floors available through salvage dealers (page 82).

Other natural materials used for hard floors include bamboo, cork and rubber, all of which are harvested without destroying the forest that produces them. Another robust choice is linoleum, made from linseed oil. Bamboo and rubber are imported from Asia and linoleum is very energy-intensive to produce, but they are all far greener than vinyl which is the product of a toxic and wasteful manufacturing process. If you are interested in rubber flooring, take care to avoid petrochemical substitutes.

Carpets

Most carpets are made from nylon, polyester and polypropylene: petrochemical-derived materials that, like vinyl, are toxic and energy-intensive to manufacture. The green alternatives are natural materials which require minimal processing, or recycled materials.

Wool carpets are a good choice, though the wool is usually coated in chemical dyes and fire retardants. There are undyed carpets on the market, and even fire retardants can be avoided, but this will narrow your options considerably. Alternatively, opt for natural materials such as sisal, coir, jute and seagrass. Their only downside is that, like most untreated carpets, they are imported.

Whatever carpet you choose, make sure the underlay is made from natural or recycled materials such as hessian, recycled tyres or natural rubber.

Costs

The cost of underfloor insulation will depend on the material you use (page 59). Choose a thickness to match the depth of your joists. Expect to pay between £2 and £10 per square metre for 100mm thick insulation.

The price per square metre for hard floor coverings ranges from £20 for bamboo and linoleum to £40 for natural rubber, all the way up to £80 for the best English slate.

Carpets start at £10 per square metre for coir, £20 for sisal and £30–£40 for untreated wool.

Further information

Talk to an energy efficiency advisor before you install floor insulation. Call the **Energy Saving Trust**'s national helpline on 0800 512 012 to be put in touch with a local advisor. The Trust's website also offers useful information on floor insulation (www.energysavingtrust.org.uk) and its free online library of best practice guides includes the *Domestic Energy Primer – an introduction to energy efficiency in existing homes*.

To find an insulation installer, contact the **National Insulation Association** (www.nationalinsulationassociation.org.uk, 01525 383313).

For up-to-date information about flooring materials and the range of products available, see the **GreenSpec** website (www.greenspec.co.uk).

The **Forestry Stewardship Council** has an online database of sustainably sourced wood products (www.fsc-uk.org).

Fridges & freezers

Fridges and freezers are heat pumps: they pump heat from the inside of the cabinet to the outside. This is why you can sometimes feel a stream of warm air at the back of your fridge. Although fridges are no longer made with ozone-depleting CFCs, they are rarely switched off so their electricity consumption and related carbon emissions continually clock up. If you want to green up your kitchen cooling, consider how you can make this heat pump do less work, less often.

Size matters most

If you are looking for a new fridge or freezer and you want to keep electricity consumption to a minimum, choose a small model. The bigger the fridge, the more energy is needed to pump the heat out. Given that heat rushes in whenever you open the door, big fridges can spend a lot of time pumping heat.

The size of fridges has grown with our dependence on supermarkets. If your life is organised around the weekly big shop, consider shopping more frequently for less. A switch to twice weekly shopping significantly reduces the fridge space you need, because perishable foods get eaten quickly and do not have to be kept for long. This is not a good idea, however, if your trip to the shops involves a long drive.

The energy label

The energy efficiency of fridges and freezers has improved since the introduction of the A to G energy label, and A-rated models can now be found everywhere. Unfortunately, however, the scheme is misleading as an A-rating is no longer the best for cold appliances. When the energy efficiency standards that define the label were tightened in 2003, two new ratings were added above A. So now you cannot buy a D, E, F or G rated fridge, but you can buy an A+ or an A++ model. The 'Energy Saving Recommended' label is now only given to A+ and A++ fridges and freezers.

To further complicate matters, the energy rating of fridges and freezers favours the bigger models. This is because the rating is based on the energy required to keep the inside cool per cubic metre. Every time you increase the size of a fridge, it is that much easier to cool the extra space you have added as you are only asking the heat pump do a bit more work on top of its core load. So energy efficiency appears to improve. But a bit more work is still a bit more work, using a bit more energy. Mega-fridges will never be green choices, however impressive their energy efficiency.

When you compare energy labels between different models, focus on the total annual energy consumption, given in kilowatt hours (kWh). This is affected by both energy efficiency and size, so it is the best single measure of the running cost and environmental impact of the fridge.

Cool spots

Many old houses have larders that are designed to stay cool passively. They are usually built on the cold side of the building and have heavy walls with small windows. If you have such a naturally cool room in your house, you might be able to forgo a fridge altogether. You could at least consider reducing the size of your fridge and keep only regularly used foodstuffs in it.

Try to keep fridges and freezers away from sources of heat such as direct sunlight, ovens, radiators and hot water pipes, and make sure there is space at the back of the fridge for ventilation to carry away the heat pumped out of the cabinet.

Costs

You should not have to pay over the odds for an 'Energy Saving Recommended' fridge or freezer, as there are plenty on the market. However, consider paying extra for a model from a top-end brand as this should offer greater reliability and durability.

Expect to pay £300 for a high quality A++ under-counter fridge.

Further information

The **Energy Saving Trust**'s database of Energy Saving Recommended products includes fridges and freezers (www.energysavingtrust.org.uk, 0800 512 012). In shops, the products included in this database will display the blue Energy Saving Recommended label as well as the standard energy label.

Furniture

It is possible to furnish a home very quickly and very cheaply. It is also possible to spend a lifetime getting the details of interior design just right. Wherever you stand between these extremes, the prospect of limiting your furniture and fabric choices solely to environmentally friendly options may seem unattractive. But don't be put off too quickly: the range of choices available is much wider than you might at first suspect. And you may even save some money.

Second-hand abundance

You don't have to think twice about the environmental impact of your shopping choices when you buy second-hand. If you buy a table that someone else no longer wants, no new resources have to be extracted from the environment, no energy is required for manufacture and no waste or pollutants are produced. You are also preventing a valuable resource from being thrown away and keeping it productive for many more years.

The second-hand market for furniture is enormous. Vintage shops, antiques markets and local newspaper listings are now complemented by furniture recycling schemes and websites such as eBay where you can buy second-hand furniture and fabrics from any period. An hour spent browsing eBay will give you more furniture choices than a week's worth of visits to department stores.

Furniture is one of the very best things to buy second-hand. If a piece of furniture has stood the test of time, you can have greater confidence in its ongoing durability. Good quality furniture will keep its character and function for many years.

Green commitment

If you are looking for new furniture or fabrics and want to stay green, look for shops and manufacturers which boast a clear commitment to good environmental practice. There are, for example, many small furniture makers working with reclaimed or sustainably sourced timber.

Unfortunately many furniture retailers will tell you that their products are environmentally friendly, regardless of whether this is true or not. If you are unsure about a company's claims, ask for details of how they source and manufacture their products. Any company with a genuine commitment to good environmental practice should be happy to provide these.

Smart shopping

When you head out into the marketplace, keep a few green questions in your mind when weighing up the pros and cons of different products:

- How long will it last? The longer a product lasts, the longer you can put off the day when more resources and energy are spent making yet another new thing.

- What is it made of? Products that are made from natural and unprocessed materials such as solid wood and bamboo have lower energy and pollution costs than synthetic and processed materials such as plastic and MDF. Some designers also work with recycled or salvaged materials.

- Were natural raw materials cultivated sustainably? Natural materials are not good eco-choices if the environment in which they are grown is damaged or depleted. For tropical hardwoods, this means only buying furniture with independent certification of a sustainable source (page 132). For cotton, this means choosing organic cotton wherever you can.

- Where did it come from? The more local the source, the lower the energy costs of transport. These can be significant for large and heavy items of furniture.

You may not succeed in answering all these questions satisfactorily every time you go shopping, but at the very least they should steer you away from really disastrous purchases.

Costs

Second-hand and reconditioned furniture can be very cheap indeed. It can also be very expensive, once you start exploring the vintage and antique markets. But there are always great bargains to be found. Similarly, new 'eco' furniture can be expensive but is not necessarily so. Much depends on the materials used.

Further information

There are many local charitable organisations involved in furniture renovation and reuse. Contact the **Furniture Reuse Network** (www.frn.org.uk, 0117 954 3571) for details.

Online furniture retailers include www.ecoinspiration.co.uk, www.thegreenhaus.co.uk and www.greenandeasy.co.uk. Furniture makers include www.ecofriendlyfurniture.co.uk, www.arborvetum.co.uk and www.ethicalwoodfurnishings.co.uk.

Organic cotton fabrics, bed linen and towels are available through many retailers including **GreenFibres** (www.greenfibres.com, 01803 868001), the **Natural Collection** (www.naturalcollection.com, 0845 3677 001) and **So Organic** (www.soorganic.com, 0800 169 2579).

Green roofs

The idea of bringing a roof to life is beguiling. Why cover a roof with sterile tiles when it could be a wild flower meadow in the sky, a joy to the eye and a rich habitat for insects and birds alike? You don't even have to have green fingers to enjoy a green roof. The simplest can be left to thrive without your help, coping with wind, frost and drought and continuing to bloom throughout the year.

Roof gardens and cliff faces

Green roofs do a lot of useful jobs. They help to keep the building cool and quiet. They protect the roof structure from weather damage. They slow the flow of rainwater into the drains, reducing the risk of flash floods. They attract wildlife and look great. Even the smallest green roof will have a positive impact on the environment in which you live.

The most substantial green roofs are roof gardens with grass, flowers, shrubs and even trees. These inevitably need a lot of soil and a very strong roof structure beneath them to support the soil, the plants and the rainwater that the soil absorbs. These 'intensive' green roofs require very careful design and installation, plus regular maintenance, so get good professional advice if you are this ambitious.

A simpler, cheaper and altogether easier alternative is a shallow, or 'extensive', green roof that you can install on an existing roof. A shallow green roof cannot support the same range of plants as an intensive roof, but there are many natural habitats, such as cliff faces, where flowers and plants survive despite poor soil and extreme weather conditions. Shallow green roofs are usually planted with sedums and other succulents that are able to store water in their flesh and so survive long periods without rain.

The right roof

Green roofs are normally installed on flat or gently sloping surfaces. However, you can exploit even a 20° pitched roof using a sedum mat: a ready-grown roll of plants that is laid out on the prepared roof surface and secured in place. Alternatively, a pitched roof can be carefully prepared with horizontal battens to prevent the soil sliding off.

The obvious places to install a shallow green roof are garages, sheds, porches and extensions. Of these, special care is needed with sheds as these are often very basic structures which may not be able to take the weight of even the simplest green roof, especially after rain. The lightest green roof is likely to weigh around 100kg/m^2 when saturated, so always seek advice to ensure the building is strong enough to support the extra weight.

You should of course be confident that your roof does not leak before you start putting a flower bed on top of it. You may want to add a further waterproof layer as part of the green roof installation, to be on the safe side.

Plant and roof preservation

The growing medium for a shallow green roof must be carefully formulated to mimic a harsh natural environment. Crushed rubble with a little soil mixed in (10%) can be used, though care must be taken not to damage the waterproof layer. You can then seed the roof or dig in plant plugs across the roof area. A sedum mat comes with a thin layer of its own growing medium but is normally rolled out onto a further loose layer on the roof.

Beneath the growing medium several membranes keep the plants thriving and the house protected. These include a layer of absorbent material such as fleece that helps to retain water, a drainage layer, a layer designed to protect the roof below from root penetration, and the all-important waterproof layer. If you are installing a green roof on top of an occupied part of your home, take the opportunity to add a layer of insulation beneath the waterproof layer to reduce your energy losses from below.

Different green roof suppliers offer slightly different approaches to the design of the green roof sandwich, so make sure you are clear what each layer is for. If you want to do the job yourself, make sure you have the right build-up to support the amount of growing medium you want. The deeper the growing medium, the greater the variety of plants you can grow – and the greater the likelihood that they will need attention once established. If your growing medium is no deeper than 10cm, it should be maintenance-free, though you may need to do some watering while it gets established.

Costs

The more complicated your roof, the more costly a green roof will be. If you have a roof that is flat or has a very shallow pitch and is free from skylights and other obstacles, expect to pay around £100 per square metre for design, supply and installation of an extensive green roof.

Blackdown Horticultural Consultants offer a green roof kit for flat or shallow-pitched roofs of 30 square metres or less. This is a practical and cheap option for small roofs. For larger projects, you will need a bespoke service (www.greenroof.co.uk, 01460 234582).

Further information

The booklet *Living Roofs* is available free on the **Natural England** website (www.naturalengland.org.uk).

For an excellent independent source of information and advice, see the **Living Roofs** website, www.livingroofs.org.

Planting Green Roofs and Living Walls (Nigel Dunnet & Noel Kingsbury, Timber Press) is a thorough guide.

Heat pumps

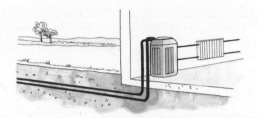

Heat pumps pump heat. It's as simple as that. They extract energy from one place, which gets cooler, and pump it to another place, making it warmer. These clever machines make it possible to take free heat from the environment and put it exactly where you want it, such as your living room. However they are machines and, like all machines, heat pumps need energy to function – and this energy isn't free.

Tried and tested technology

You probably own a heat pump already: a fridge. A fridge removes energy from the cold interior, cooling it down further, and releases it out the back, creating a stream of warm air. 'Ground source heat pumps' work in exactly the same way but their purpose is to provide warmth rather than to cool things down. They extract energy from the ground, where the temperature is a steady 10°C, and transfer it to the inside of your house. The sun beats down on the ground all year, so there is no shortage of energy stored in the earth. All you have to do is pump it out. Heat pumps do not store heat, they just move it from one place to another.

Three independent loops of fluid are needed to perform the heat pump's task. The first is a loop of antifreeze that runs under your lawn and absorbs heat from the ground. The last is the loop of hot water that emits this heat into your home through underfloor heating or radiators. In between is the loop of refrigerant inside the heat pump that shifts the energy between the first and last loop. As this loop is shifting energy from a relatively cool fluid to a warm fluid, it cannot work by the natural flow of energy. Instead, the refrigerant is repeatedly evaporated, compressed and condensed to pump the heat in the 'wrong' direction.

The catch

Quite a lot of electrical energy is needed to power the refrigerant cycle, especially the compressor. This means that heat pumps are not necessarily cheap to run or 100% green, despite the fact that they are extracting spare solar heat from the ground. Electricity is the most expensive domestic fuel and has the highest carbon emissions because so much energy is wasted in the power station (page 32). Electricity has about 2.5 times the carbon emissions of natural gas (the cleanest fossil fuel) and costs about three times as much. However the heat energy provided by a ground source heat pump is typically three to five times greater than the electrical energy needed to run it, so heat pumps are still good eco-choices compared to gas, LPG, oil or direct electric heating.

The right house for a heat pump

Heat pumps are best suited to well-insulated, energy-efficient homes. Unlike boilers, they cannot produce a huge surge of energy to heat up a house quickly but prefer to chug away in the background, providing a regular output of heat to keep the building at a stable temperature. If you live in a house that cools quickly and is often unoccupied during the day, a heat pump will not be your best option.

Heat pumps become less efficient as the difference between their source and output temperatures increases. Consequently they are ideal for underfloor heating which operates at a lower temperature than radiators. To enable the heat pump to work efficiently, radiator systems may have to be upgraded with bigger radiators which emit more heat.

Sizing the pump and ground loop

Heat pumps come in many sizes and can supply large houses if necessary. Modern models can provide your hot water too. If you are thinking of installing a heat pump, get an independent energy survey of your home done first in order to specify a model and ground loop big enough to cope with your heating demand. Wherever you can, reduce this demand first with improved insulation and draught-proofing. Be wary of heat pump suppliers who attempt to calculate the size of your heat pump and ground loop pump based on rules of thumb. It is vital that your ground loop is big enough to meet your needs.

Ground works

The ground works for a heat pump can be substantial. Depending on the space available, you can either dig long trenches for the pipework in your garden, at about 2m deep, or put the pipes down a 75-100m borehole. Get a soil survey done first to ensure that the ground is suitable. The trench option is easier and cheaper but may seriously mess up your garden during installation.

Some heat pumps do not need ground works: air source heat pumps extract heat from the ambient external air or, in some models, from the exhaust of a mechanical ventilation system.

Unlike the ground, which stays at a stable temperature all year, the air gets very cold just when you want to extract the most heat from it, so the efficiency of air source heat pumps can be very poor. They should only be considered for small properties with very low energy demands.

It is also possible to draw heat from water such as a flowing river. This can be a good heat source because the extracted heat is rapidly replaced. However, you must get permission from the Environment Agency before you start extracting heat from a water course.

Suppliers and quotes

Do not expect a heat pump supplier to provide everything you need. If you plan to upgrade your central heating with underfloor heating or bigger radiators, you will need to organise this independently. This may mean that you have to co-ordinate different plumbers and electricians at different times.

A quote for a ground source heat pump will include some or all of the following: site investigations, system design and specification, the heat pump itself, the external pipe work, the ground works, installation, commissioning, and a service agreement or guarantee. Do not be surprised if you get very different quotes from different companies. Check to see what is included in every quote and ask the suppliers to explain their quotes in detail.

Costs

The cost of a ground source heat pump depends on how much heat it needs to pump, which in turn depends on the size of your house and how much heat you lose. A well-insulated medium-sized house will typically need a heat pump that can supply between 6kW and 10kW of heat. This will cost between £5,000 and £12,000 to supply and install, excluding the heat distribution system within your home. You will be looking at the higher end for ground works involving boreholes.

Grants are available to support the installation of ground source heat pumps. Contact the Energy Saving Trust to find out what you are eligible for (www.energysavingtrust.org.uk, 0800 512 012). If you apply for a grant, you have to use an installer registered with the scheme.

Further information

Useful information and supplier details are available from the **Ground Source Heat Pump Association** (www.gshp.org.uk, 01908 665555).

The **Energy Saving Trust**'s free online library of best practice guides includes the *Ground Source Heat Pumps Factsheet, Heat Pumps in the UK: a monitoring report* and *Domestic Ground Source Heat Pumps: Design and installation of closed-loop systems* (www.energysavingtrust.org.uk).

A good online guide to heat pumps is provided by **John Cantor Heat Pumps** (www.heatpumps.co.uk).

Insulation

Three-fifths of the energy we burn in our homes goes on heating, but most of this energy rushes straight out of our homes to heat the streets and sky. We think that the purpose of our central heating systems is to keep us warm, but they merely compensate for the failure of our buildings to do this job. A dwelling ought to keep the rain off and the warmth in. If you want your home to achieve the latter of these basic tasks, buy some insulation.

You can't be green without it

If your home is short of insulation, make this your number one greening up priority. The benefits include increased warmth and comfort in the winter, a cooler interior in the summer and a quieter home throughout the year. You will also enjoy greatly reduced heating bills and your carbon dioxide emissions will tumble.

The easiest place to start is your loft, if you have one. But don't stop there: your walls and floors can also be fully insulated. For details of the practicalities of installing insulation, see the entries on Roofs and lofts (page 76), Floors and floor coverings (page 37) and Walls (page 111).

There are many different types of insulation on the market. These include inorganic mineral fibres such as rockwool or glassfibre;

petrochemical-derived foams, boards and beads; and natural organic materials such as sheep's wool, hemp, flax and cellulose. As almost any insulation is a green purchase, you don't have to worry too much about which one to choose, but consider the following issues first.

Filling the gaps and staying put

Insulation will not work properly if it does not fill its allotted space, as heat will flow quickly through any gaps. It is therefore essential that the material you use has the flexibility or rigidity to fully fill a space and stay there.

Insulation laid between joists in a loft space needs to be flexible to fill the space, so mineral wools and natural organic materials are often used here. But rigid boards, typically made from polystyrene or other petrochemical products, are useful when the insulation needs to be self-supporting, such as for internal wall insulation or concrete floors.

Some spaces can only be reached by blowing the insulation in. For cavity walls, fragments of mineral wool or polystyrene beads are used to fill all the hidden spaces in the wall.

The space available

The rate at which heat flows through insulation varies considerably between different products, so for a given space some types of insulation will do a much better job of keeping the heat in than others. Petrochemical-derived materials such as phenolic and polyisocyanurate boards lead the field. To compare products, look for the measure of thermal conductivity: the lower the conductivity, the more effective the insulation. Sometimes the thermal resistance value is given instead, which increases with better performance.

This is an important issue if the space available for insulation is very limited. For example, interior wall insulation encroaches on the space you live in, so dry lining products that combine plasterboard with insulation are typically made with high performance petrochemical-derived products.

Handling

Some insulation materials are not very pleasant to use. In particular, mineral wool can irritate both the skin and the lungs, so you should definitely wear a mask and gloves when installing this. Some mineral wool products are encapsulated in foil to make them easier to use. This is where natural organic materials come into their own: sheep's wool, for example, is a pleasure to use.

Moisture

Insulation can affect the way that moisture moves through the fabric of your house, and there is sometimes a risk that water vapour will condense on the cold outer side of insulation within a wall or roof structure, which could severely damage your home. This is why your loft should be well ventilated above the insulation layer to remove any condensation that may form in the loft space.

Some insulation materials cope with moisture much better than others. Natural organic materials do well: cellulose (including recycled newspapers and straw), sheep's wool, hemp and flax absorb and release moisture as the outside temperature cools and warms over the day.

Wider environmental impacts

All insulation materials do a great job of reducing the environmental impact of your home, but some materials also boast very low environmental impacts during manufacture. Natural, minimally processed materials such as hemp, flax and sheep's wool do well here, as do recycled materials such as newspapers and fabrics.

Many petrochemical-derived materials used to be made with blowing agents that were very harmful to the ozone layer. If you are buying one of these products, make sure it is classified as having 'zero ozone-depletion potential' (zero ODP).

Costs

There are big variations in the prices of insulation. The cheapest materials are the mineral and glass wools. You may have to pay considerably more for natural organic materials or some high performance petrochemical-derived materials. Never compromise on the potential performance of the installed insulation by buying something that is cheap but inappropriate.

For example, the price per square metre of 100mm depth of loft insulation is around £2 for rockwool, £2.70 for pulped recycled newspapers (Warmcel 100) and £8.50 for flax (Isovlas).

Grants for insulation may be available through local programmes or your energy supplier. Contact the Energy Saving Trust to find out what you are eligible for (see below).

Further information

Call the **Energy Saving Trust**'s national helpline on 0800 512 012 before you install new insulation, to be put in touch with a local advisor. Their website also offers useful information on insulation (www.energysavingtrust.org.uk) and its free online library of best practice guides includes the *Insulation materials chart – thermal properties and environmental ratings* and the *Domestic energy primer – an introduction to energy efficiency in existing homes*.

For up-to-date information about insulation materials and products, see the **GreenSpec** website (www.greenspec.co.uk).

Lighting

Most of the lighting we use in our homes is provided free by the sun. If you have to use any other light source while the sun is still up, your windows are not doing their job. Consider what you can do to bring more daylight in and so keep your electric lights turned off. Then ask yourself if your lights and lamps provide light where it is really needed. Last, but certainly not least, change your light bulbs.

Daylight

The easiest way to improve the penetration of daylight into your home is to paint your walls and ceilings white or a very light shade of colour. Keep windows and windowsills clear and use blinds that will reduce glare without too much loss of light. Try to locate regular tasks near the daylight: a kitchen sink is always best placed under a window.

Adding more glazing to your home will of course bring in more daylight. For their size, roof windows (skylights) are particularly effective in doing this. If you want to bring daylight into a dark corner but a window or roof window is impossible or too costly, you may be able to channel light from a rooftop using a light tube, a glass bauble that gathers daylight on the roof and sends it down a highly reflective tube to the room below. These are small enough to fit between rafters and joists, avoiding the need for major structural works.

Lighting design

Good lighting design can transform your experience of your home. If you are used to bland, brightly lit interiors, try to rethink your lighting based on how you use each room. Where is light needed for tasks such as cooking or reading? Where would light look good as a feature? And how much general ambient lighting do you really need? Often we use too much ambient lighting, flooding every corner of every room with light, and live with poor task and feature lighting. Banks of ceiling-recessed halogens are the worst offenders.

Light bulbs

Traditional tungsten incandescent lightbulbs have barely changed since they were invented by Thomas Edison over a hundred years ago. They are phenomenally inefficient and will soon be banned by the European Union. Small tungsten halogens are only marginally better: although they are often low voltage and run off transformers, this does not mean they have low power consumption (50W is common).

Compact fluorescent 'energy-saving' bulbs typically use about a quarter of the electricity of the equivalent incandescent bulb (e.g. 9W instead of 40W). Although old-fashioned fluorescent tubes give off a cold blue light, many modern compact fluorescent bulbs emit a warm light and can replace an incandescent bulb without any change in mood. They also come in all shapes and sizes including spotlights, candle bulbs and ceiling-recessed downlighters. All compact fluorescent bulbs run off mains voltage so you may have to change your ceiling fittings as well as your bulbs if you have low voltage downlighters. Look for GU10 fittings and bulbs.

Many compact fluorescent bulbs take a short time to warm up to full brightness, which takes a little getting used to. But do not leave them on unnecessarily – the idea that fluorescent bulbs should be left on to save energy has always been a fallacy. Most compact fluorescent bulbs cannot be dimmed but special dimmable versions are now available.

The rising stars of low energy lighting are light-emitting diodes (LEDs). The light quality of LEDs is sometimes disappointing so experiment before you buy in bulk. Currently they are best used as small feature lights, including garden lights, rather than for ambient or task lighting.

Costs

The most energy-efficient roof windows start at £400 plus installation (www.velux.co.uk). Light tubes cost a minimum of £200 plus installation (www.sunpipe.co.uk). Rigid tubes channel the light much more effectively than flexible tubes.

Good quality compact fluorescent bulbs cost more than incandescent bulbs but they last several times longer and use a fraction of the electricity so their lifetime cost is much lower.

You need to look beyond the supermarkets and DIY stores for the full range of compact fluorescent and LED lights. Try online retailers such as www.ecoelectricals.co.uk and www.greenstock.co.uk.

Painting & decorating

The urge to redecorate comes to us all. The next time you feel it, resist the temptation to nip down to the DIY shop for a tin of processed petrochemicals and investigate the many green alternatives instead. Modern eco-paints will give you an attractive and durable finish without leaving a trail of toxic waste behind them.

Paints and finishes

Paints and finishes made from plants and minerals are natural products, so they cause little harm to the environment in manufacture, require far less energy to produce than petrochemical paints, and can be safely returned to the natural world without polluting it. Some of the natural paint makers even suggest you add the dried-out leftovers of their paint to your compost heap. In contrast, petrochemical paints are very difficult to dispose of without causing pollution.

Any tin of paint, stain or varnish with a 'high VOC' label on it should be avoided as its contents will release lots of toxic volatile organic compounds into the air as it dries. Unfortunately 'low VOC' petrochemical paints and finishes, including water-based paints, still contain many other potentially harmful chemicals. In contrast, natural paints will not leave you with a headache. A few natural ingredients are also classed as volatile organic compounds, such as turpentine and citrus oils, but even these can be avoided if you use natural water-based paints. Unlike petrochemical paints, natural paints usually come with a full list of ingredients, so you can check out what you are breathing.

Using natural paints

As natural paints are completely different products from petrochemical paints, don't assume that they will be the same in use. They smell different (some sweet, some stinky) and some are sloppier to apply. If you are doing a lot of painting, you might want to experiment first with different brands. If you are getting the painters in, be prepared for them to complain about using paints they are not familiar with.

Unlike petrochemical paints, which create a plastic film on your walls, natural paints are porous. This means that any moisture beneath the paint can be released without the finish flaking or cracking. Walls finished with lime render must be painted with porous paints. If you have any concerns about damp in your walls, avoid petrochemical paints as they may make the problems worse.

The simplest natural paints are limewashes, distempers and casein paints. You can even make these yourself from the basic ingredients. However, you may need to paint many coats to achieve a result you are happy with, and they are not as durable as commercial natural paints. Although natural paints come in a wide range of colours, you can also create your own colours by adding mineral powders and other natural pigments to white paints.

Wallpaper

Vinyl 'wallpapers' are not papers but plastics, made from a cocktail of petrochemicals. However, much of the paper used to make real wallpapers comes from virgin forest, so the only way you can be sure that you are buying real paper from a sustainable source is if it comes with the FSC (Forestry Stewardship Council) logo (page 132) or if it is recycled. Very simple papers, such as lining papers, are relatively easy to find in DIY stores with this provenance. Designer papers are harder to find, but it's always worth asking.

Tiling

Your best green option for tiling is a locally sourced unfired material such as stone. You may need to stretch your definition of 'local' to find what you want, but do your best to avoid materials transported from the other side of the world. Although ceramic tiles have a higher energy cost than stone because they are fired, they are very durable, so this cost is small across a long lifetime of use. Focus on buying tiles which you are confident will not be swept away with next year's change in fashion. If your tiles need a final finish, there are natural products available, usually based on linseed oil.

Limited choice?

Eco paints do not come in the seemingly infinite range of colours that synthetic manufacture has made possible, and you will reduce your options if you only consider wallpaper which you know has been sustainably sourced. But don't let this put you off going green. The least you can do is explore the green options and give them priority when the colours work for you. The best you can do is work creatively with a still extensive palette based on natural materials and pigments.

Costs

As natural paints are not a convenient by-product of the petrochemical industry but are made from specially sourced natural materials, they tend to cost more. At the top of the market, five litres of Auro white emulsion costs £31, about twice the petrochemical alternative (see www.auro.co.uk for stockists). Other brands of natural paints and finishes available in the UK include Aglaia, Biofa, Earthborn, Holkham and Livos. Osmo specialises in wood finishes. These paints and finishes are rarely available in ordinary DIY sheds but can be bought from specialist suppliers online (page 141) and independent shops.

Home Strip is a non-toxic alternative to caustic paint and varnish removers. Half a litre costs £8 (see www.ecosolutions.co.uk for stockists).

Further information

Using Natural Finishes (Adam Weismann & Katy Bryce, Green Books) is a comprehensive, fully illustrated guide to making and using all kinds of lime- and earth-based plasters, renders and finishes.

See also *The Natural Paint Book* (Lynn Edwards and Julia Lawless, Kyle Cathie).

Volume 2 of *The Green Building Handbook* (Tom Woolley and Sam Kimmins, E & F N Spon) contains detailed analysis of the contents and effects of petrochemical and natural paints.

Paving

Cities do not have to be concrete jungles: with a little attention to detail, nature can thrive alongside our hard-faced habitats. This is not just a matter of protecting gardens but of treading lightly upon the land in every way. Paving materials are important because a vast area of urban land is covered by domestic driveways and patios. If you are planning new surfaces on which to tread or drive, go green and soften up.

Soak it in

A concrete or tarmac drive has an impermeable surface which sheds the rain straight into the storm drain. When everyone's drives are impermeable (as well as all the roads and roofs), the pressure on the drains is considerable and can lead to flash floods. Most storm drains empty into water courses which can be damaged by sudden rapid flows of often polluted water. If your storm drain is connected to a sewer, every drop of rainwater that you send down the drain adds to the burden on the water treatment system.

If you use permeable paving, the rain will soak into the ground rather than being flushed into the drains. This reduces the pressure on water courses and water treatment plants, helps to filter out pollutants, cuts the risk of flooding, sustains local plant life and ultimately contributes to the replenishment of aquifers. Maintaining life and growth in your front garden also makes a big difference to the quality of the street environment.

Hard and soft

Many permeable paving systems combine a hard structure to provide support with a soft, porous medium to absorb the rain. These include concrete pavers in an open lattice and interlocking cellular paving systems made from recycled plastic. These cellular systems are designed to hold a top-dressing of turf, gravel, stone chips or recycled aggregate such as crushed glass. Once filled, the plastic cells are barely visible yet remain strong enough to support a vehicle. The cells protect the top-dressing and prevent turf turning to mud or stones getting scattered.

You can create permeable paving with ordinary hard paving materials simply by leaving gaps between them. You only need to pave the ground that the wheels of your car travel over, not the entire drive.

If you want the look of a fully paved drive, use paving products designed specifically for this purpose which channel water through or between the pavers.

Let the grass grow – or not

The use of permeable paving is an opportunity to create a beautiful low-growing garden between or through the pavers. Plants such as creeping jenny and bugle will even cope with regular pressure from the wheels of your car. This approach does, however, require regular attention, not least to remove the dandelions and other weeds that will threaten to crowd out your alpines. A water-permeable weed suppressing fabric (available from garden centres) installed beneath the paving will give you a low maintenance permeable drive.

Material choices

Reclaimed and recycled materials are the best choices for pavers or top dressing. York stone and other heavy pavers are widely available through salvage yards – always the best eco choice because no new resources or energy are expended. For suppliers of recycled paving materials and aggregates, including recycled glass, see www.recycledproducts.org.uk.

For new materials, your best choices are natural, local and unfired such as wood chip, British stone and locally quarried gravel. Try to avoid concrete (including fake stone pavers) and imported stone.

Costs

Grassington recycled plastic cellular paving costs around £12 per square metre (www.greenwaydirect.co.uk). Blockley's Hydrosmart permeable clay pavers start at £13 per square metre (www.blockleys.com).

The price of reclaimed York stone is similar to new stone, with decent quality stone starting from around £50 per square metre. Garden woodchips are very cheap (less than £10 for 100 litres) but must be renewed every couple of years.

Further information

For an overview of all forms of paving, including permeable and reclaimed paving, see the **Paving Expert** website, www.pavingexpert.com.

Gardening Matters: Front Gardens, the **Royal Horticultural Society**'s guide to combining parking with gardening, is available free on their website (www.rhs.org.uk).

Rainwater collection

Water is a precious resource. Happily it is also a renewable resource: the rain clouds just keep rolling in. But hot summers, dry winters and ever-increasing demand are creating serious water scarcity. It therefore makes good green sense to reduce your consumption of mains water in whatever ways you can. Your first step should be to use water more efficiently in the home (see pages 24, 83, 100 and 116), but rainwater collection can also make a big difference, especially if you have a garden.

Water butts

The simplest means of rainwater collection is a water butt attached to a downpipe from your gutter. Water butts are worth installing even in the smallest of gardens because the greatest stress on the water supply occurs in summer when everyone is pouring high quality drinking water onto their lawns. Rainwater is also useful outside for car washing.

Water butts are widely available in garden centres and DIY stores or at discounted prices from your council or water company. They come in many shapes and sizes beyond the basic barrel shape, including discreet boxes for fixing to walls and terracotta pots. You can even buy plumbing kits to link water butts together if your existing butt is not big enough. If you use a lot of water outside, consider installing a large tank and using a hand-held electric pump to increase the flow rate beyond that of a watering can.

Rainwater inside the home

You can use rainwater inside your home, but it must be kept separate from mains water. The amount of extra piping required depends on where your rainwater is stored and where you want to use it. Rainwater is usually only used for flushing toilets and filling the washing machine. You should stick to mains water for drinking, cooking and personal washing. If you have low-flush toilets (page 101) and a water-efficient washing machine (page 116), you will only save a moderate amount of mains water inside the home by installing a rainwater collection system.

Pumped systems

There are various commercial 'rainwater harvesting' systems, all of which employ large tanks sited outside the house, in a basement or underground. On average they save 40-50% of household's mains water consumption.

Typical tank sizes range from 2,000 litres for a small terraced house to 5,000 litres for a four bedroom family home. In most systems, the rain from the roof passes through a simple filter into the tank which, if it gets full, overflows into the stormwater drain. On demand, a pump sends the rainwater from the tank down dedicated pipes to the toilet cisterns, washing machine and garden tap. If the water in the tank gets too low, it is automatically topped up from the mains.

Rainwater harvesting systems are completely automatic and do not require much maintenance beyond keeping a regular eye on the filter. Above-ground tanks may also have to be disconnected in icy conditions.

Although rainwater harvesting systems are not always simple or cheap to install, they are worth considering if you use a lot of water both inside and outside your home. Get advice first to be sure that you have enough roof available to feed the system.

A DIY gravity-fed system

If you have space for a tank indoors somewhere below your highest gutter and above your toilet cistern you can collect rainwater directly from your roof and let it flow into the cistern when required. The tank must have a back-up mains water feed, positioned to prevent any backflow of rainwater into the mains pipe, so that the toilet keeps flushing during a dry spell. The overflow can be directed down the drain or ideally to an external water butt. If you can prevent excess rainwater going down the drain by using a ground 'soakaway' that absorbs the water and keeps the pressure off the drains, so much the better.

Costs

Basic water butts start at £25 but go up to £350 for really large tanks. Rainwater harvesting systems cost from £1,500 to £2,000 for small to average-sized houses. The installation cost including plumbing and hole-digging is likely to be £1,000 to £2,000.

Further information

The **UK Rainwater Harvesting Association** provides information and supplier contacts (www.ukrha.org).

A free booklet on *Conserving Water in Buildings* is available from the **Environment Agency**. This includes a section on Rainwater harvesting (www.environment-agency.gov.uk, 08708 506 506).

Recycling

Recycling saves energy, water and raw materials, and reduces pollution and waste. It ought to be taken for granted, yet many people still don't do it. This may be because they don't want to do it, or don't believe in it, but it is also because their homes are not designed to make recycling an everyday, easy activity. As our houses are rarely built with recycling in mind, this is an excellent way to green up your home with some simple DIY.

Reduce, reuse, recycle

Before you start recycling, do everything you can to reduce your consumption and to reuse materials where possible. Many of the chapters in this book describe ways of reducing energy and water use in the home, and there is much you can do beyond the home – especially in shops where plastic bags and packaging still proliferate – to reduce the unwanted materials that enter your home. When you are planning home improvements, do your best not to over-buy materials which may end up being thrown away. Make the most of second-hand markets (including the online versions) when disposing of materials or possessions that you no longer need.

Waste or resources?

Modern kitchens come with extraordinary amounts of storage space, yet they rarely include anything more than one little bin for rubbish. Concern for the many resources that come into a kitchen is not matched by concern for the resources that leave it. In fact what leaves a kitchen is assumed not to be a resource but simply waste.

In practice almost everything that leaves a home can be thought of as a resource with potential for a new use through recycling. You simply have to make the space to store these materials so that you can deal with them appropriately. The kind of storage you need will depend in part on how your local council accepts recycled materials. Some councils take them all mixed up, others expect you to separate them first.

Bins, bins, bins

Think abut the different resources that you regularly generate and where their storage is best located. Organic materials from the kitchen need a bin in the kitchen. Tins, cans and plastic bottles are also best stored near to where they are used in the kitchen. Glass bottles and paper are bulkier and so may be better stored near the front door, ready for removal. You will of course still need a genuine 'waste bin' on top of all this storage, for the things you cannot recycle, such as plastics your council will not accept and packaging made from combined materials that you cannot separate.

As well as simple lidded bags attached to the inside of cupboard doors, you can buy bins that slide out and open when the cupboard opens, bins that swing into the hidden space of corner units, and bins that are actually set into the work surface with a flush lid. Assess how much you produce of any material before you choose your bin, as a bin that fills too quickly and overflows will soon drive you to distraction. However, keep organic waste bins small so that composting does not start before the scraps have left the kitchen.

Composting

Almost all kitchen waste can be composted. Fruit and vegetable peelings mixed up with cardboard will slowly turn into a rich mulch in an ordinary compost bin in the corner of your garden. If you only have a back yard or you want to compost your cooked food waste too (which attracts vermin in an ordinary bin), invest in an insulated tumbling bin, the quickest of all composting methods. Cheaper options for cooked food are bokashi bins and green cones.

Recycled products

Sending materials to be recycled is of little value if there is nobody willing to buy the material and make it into something new. Do what you can to encourage the market for recycled products wherever possible. Go beyond bin bags and toilet paper to crockery, compost and coffins (see www.recycledproducts.org.uk). A database of recycled building products can be searched at www.ecoconstruction.org.

Further information

Reduce, Reuse, Recycle (Nicky Scott, Green Books) is a comprehensive domestic A to Z.

The national recycling website is www.recycle-more.co.uk. To find out where you can recycle everything beyond what your kerbside scheme will take away, see www.wasteconnect.co.uk.

The Bin Company (www.thebincompany.com) has a wide range of bins.

Composting: an Easy Household Guide (Nicky Scott, Green Books) is a good introduction. See also www.gardenorganic.org.uk. For a variety of composting equipment, see www.wigglywigglers.co.uk, www.smartsoil.co.uk, www.livingsoil.co.uk, www.greencone.co.uk, and www.bubblehouseworms.com.

Roofs & lofts

Roofs are the most important and most forgettable part of our homes: we want a roof to be robust and long-lasting, quietly doing its job without our intervention. But don't ignore your roof if you want to green up your home. Whether or not you are interested in the material that keeps the rain off, make sure that the material that lies below, keeping the heat in, is as good as it possibly can be.

Insulating your loft

Heat rises, so insulating your loft is a very effective way of cutting your heat losses and reducing your carbon emissions. As heating accounts for 60% of the energy we use in our homes, anything you do to cut your demand will make a difference – and loft insulation will make a big difference. If you already have some insulation in your loft, put some more on top. The minimum recommended depth is 270mm.

Rolling out insulation between the joists is relatively easy but ideally you should also put a layer of insulation on top, at right angles to the first layer, covering the joists. This can be done with the same insulation but you will need to fix supports first for any final boarding out, if this is needed. Alternatively you can buy rigid insulation boards to put on top of the joists. For details of the range of insulation materials available, see page 56.

Be sure not to block up the air gap at the eaves (the bottom of the roof slope), so that the cold interior above the insulation remains properly ventilated and any condensation is cleared. It is worth laying plywood or hardboard sheets at this edge to maintain the air gap and to ensure that the cold air doesn't get under the insulation.

Your goal is to create a continuous warm blanket over the entire roof space. Do not leave any gaps in the blanket through which heat can escape. If you have a loft hatch, this should be insulated and the junction with the ceiling fitted with good quality draught-strip (it may be easier to fit a new fully insulated door). If you have a cold water tank in your loft, remove it or thoroughly insulate it, or else it will freeze in the winter. Any pipes in the loft space should also be insulated with purpose-made pipe insulation.

Insulating your roof

An alternative to insulating the loft floor is to insulate the roof, filling and boarding over the sloping rafters in the same way that you would fill and cover the joists. This is essential if you are having the space converted into a habitable room – a loft conversion is a great opportunity to maximise the warmth of your new walls and roof. As the depth of insulation will not match what you can achieve on the floor, you may have to spend more money on higher performance insulation to get the same result. You must also include thorough draught exclusion or all the heat in your new warm loft will just blow away. Ventilation is still needed, but on the cold side of the insulation, so a gap must be left between the insulation and the roof felt, extending down to the gap at the eaves.

If you are renovating or replacing your roof tiles, insulation can be added to the outside of the rafters first. This is called 'sarking' insulation and requires professional installation.

The many uses of a roof

If you have plans to renovate or improve your roof, first consider the many ecological functions that it could perform beyond keeping you warm and dry. You could get hot water from your roof using a solar hot water panel (page 91). You could generate electricity using photovoltaic cells (page 89). You could create a rich habitat for local wildlife with a roof-top meadow (page 47). Finally, you could channel all the water that lands on your roof to a collection tank from where it can be used both outside and inside your home (page 70).

Roofing materials

If you live under thatch, your roof is about as environmentally friendly as it could be, as Norfolk reed is a natural, renewable resource which takes very little energy to harvest and transport. For similar reasons, timber tiles (shingles) are a good choice. These are usually made from cedar or oak, naturally durable woods that do not require preservative treatment. Look for a British rather than North American source, and make sure they come with certification of sustainable forest management (page 132).

If you have a slate roof in need of repair or refurbishment, try to source your slates from Wales rather than China. The price will be higher, but the slates will last longer and carbon emissions from transport are much lower. Good quality clay and concrete tiles also score well for durability, though clay tiles require a lot of energy to make and concrete manufacture is a major source of carbon dioxide emissions. If you can source any of these materials from a local salvage yard you will have made the greenest possible choice (page 141). Avoid fake tiles that are made from metal or plastic, as their manufacture is both toxic and energy-intensive.

Costs

Loft insulation is not expensive and is very cost-effective. The cheapest insulation is mineral or glass wool. If you already have 100mm insulation in your loft, adding another 170mm will cost around £3.50 per square metre. If you are starting from scratch, expect to pay at least £5.50 per square metre. Natural, organic materials such as sheep's wool, flax or hemp will cost considerably more, though loose-fill bags of Warmcel 100 recycled newspaper cost around £8 per square metre for the full depth.

Roof insulation is more expensive, but is usually installed as part of a roof renovation or loft conversion when the marginal cost of specifying high performance insulation is relatively low.

Further information

Talk to an energy efficiency advisor before you install loft or roof insulation. Call the **Energy Saving Trust's** national helpline on 0800 512 012 to be put in touch with a local advisor. The Trust's website also offers useful information on loft insulation (www.energysavingtrust.org.uk), and its free online library of best practice guides includes the *Domestic energy primer: an introduction to energy efficiency in existing homes, Energy efficient loft extensions*, and *Advanced insulation in housing refurbishment*.

For up-to-date information about roofing materials and the range of products available, see the **GreenSpec** website (www.greenspec.co.uk).

To find an insulation installer, contact the **National Insulation Association** (www.nationalinsulationassociation.org.uk, 01525 383313).

Salvage

Using salvaged materials in your home and garden is rarely the easiest option, but it is always the greenest option. The more we keep old materials in use, the less we damage the environment through the extraction, processing and transport of new materials. Reclaimed materials need transportation too, but the salvage industry is much more localised than the global market for new products. A wooden floor from the school up the road is infinitely preferable to newly imported tropical hardwood (assuming the school is happy to lose it, of course).

What's available?

Reclamation yards often present themselves as architectural treasure troves, stacked high with quirky features and unusual garden ornaments. This can be misleading, as almost any building component or household fixture can be bought second-hand. Don't just head for the local yard when you want a feature for your garden. The environmental benefits of salvaged materials really begin to register when they are used in bulk as replacements for new materials.

Salvaged materials can be used for floors, paving, roofing and brick walls. Work surfaces, sanitaryware, doors and fireplaces are also widely available. Whatever the period or style of a home, there is always scope for using salvaged materials within it. Some period fittings may only suit certain period properties, but even the most contemporary home can benefit from the character of used materials such as stone, wood and fired clay.

Quality

Used materials have already demonstrated their durability. Some may be a little worn or weathered, but this often adds to their charm and value. If you need a finish that will cope with serious long-term abuse, such as a floor, roof or pathway, salvaged materials are a good choice.

You must, however, keep a watchful eye on the quality of your purchases. Unlike new and recycled materials, salvaged materials come with no guarantee or British Standard certification. To avoid cracked bricks and oil-stained paving stones, buy from a reputable dealer and check the materials as fully as you can before confirming the purchase.

Many salvaged materials will be in good structural condition but will nonetheless require extra work to install. Always consider exactly what you will need to do to materials or fittings to bring them to life in your home. Variations in size, such as the depth of paving flags, are likely within even the most carefully sorted batch of materials, leading to problems during installation. Any cleaning and preparation of materials can be time-consuming for large surface areas, and sanitary and electrical fittings are likely to require special attention.

Reputable sources

There is a thriving market in stolen architectural features, especially garden features that are relatively easy to nick. Some reclamation dealers have signed up to the Salvo Code, a voluntary commitment not to buy goods if there is any suspicion that they may be stolen, but this is not monitored. The best you can do is stick to reputable dealers, ask where any purchase originated and, if you have any doubts, check the Salvo online database of stolen property (www.salvo.co.uk).

Costs

Some salvaged materials are more expensive than new materials, others are cheaper. Price is usually a reflection of the quality of a material, rather than whether or not it is reclaimed. There is, for example, far greater variation in the price of all timber flooring than there is between new and reclaimed floors.

For example, reclaimed oak parquet might cost you £40 per square metre compared to £30 when buying new. Reclaimed Burmese teak parquet will cost you much more, around £75 per square metre, and you will be hard pushed to find a sustainably harvested new product to compare this price with.

Further information

Salvo maintains an excellent online directory of reclamation yards in Britain (www.salvo.co.uk).

Showers, baths & taps

The easiest way to green up the time you spend in the bathroom is to get out of the bath and into the shower. But this isn't the end of the story. In fact some showers use even more water than baths. Only if you pay attention to the finer details of your bodily ablutions can you hope to be exceptionally green and clean.

Save water – save energy

Showering and bathing require two precious resources: water and energy. As water has a very high heat capacity, a lot of energy is needed to heat it. If you can reduce the amount of water you use, you will also cut your energy bills and carbon emissions.

Three things affect how much water you consume: what you wash with, how you wash and how often you wash. The last of these is a matter for you and those close to you. It is worth bearing in mind, however, that daily showering is a recent phenomenon and a daily shower may consume more water per week than a twice-weekly bath.

The right shower

On average, a shower requires 30 litres of water compared to 80 litres for a bath. Power showers, however, pump out water so fast that they can consume more water than a bath in one go. They should be avoided at all costs – you simply don't need this speed of water flow to enjoy a good shower.

Many ordinary showers have poor controls, which take time to adjust and bring to the right flow rate and temperature, wasting a lot of water before the shower really begins. Separate hot and cold taps are the worst offenders. If you are fitting a new shower, make sure it has a single temperature control and a separate flow control. Both should be sensitive and easy to adjust. Choose a thermostatic mixer which maintains the same temperature regardless of the flow rate. Many shower heads also give you some degree of control over the flow of water.

Ultra-efficient showering

Just as it is sometimes a pleasure to relax in a bath, so it can be a pleasure to stand under a shower and enjoy the water pouring over you. But if all you want to do is get clean, this is not the most efficient way of showering. With good controls on your shower, you can rinse, turn off, lather up, then rinse again. Showering in this way with brief bursts of water can cut your water consumption by 50% or more. If you turn off the tap while you are brushing your teeth, follow the same logic when you get in the shower.

Taps

If you are fitting new taps and you have mains pressure water or a combi boiler, make sure the taps have aerators fitted at their outlets which mix the water with air and so give the illusion of a full flow for less water. You can also retrofit water-saving devices to existing taps, as long as their outlets are round. The Tapmagic cartridge delivers a water-saving spray when you open the tap (ideal for brushing your teeth) then shifts to a high flow when you turn the tap on full to fill a basin.

Do not use aerators or flow restrictors for bath taps. You want these taps to run as quickly as possible so that you lose a minimum of heat before getting in.

Recycling energy and water

It is possible to pipe the warm water leaving your shower through a heat exchanger and transfer the heat to the cold water entering your heating system, but unfortunately no commercial systems are currently available in Britain. You can easily put the energy in a bath to a second use in the winter months: when you get out, don't pull the plug. Let the heat dissipate into the air and warm your home, then empty the bath in the morning when the water is at room temperature.

It might seem like a good idea to use the waste water from showers and baths – grey water – to flush your toilet but unfortunately this water is full of bacteria, which rapidly multiply when the water is stored, and so starts to stink in no time. Grey water is best reused immediately on your rose garden, if you can get it there, or not at all.

Costs

You won't take showers instead of baths if you don't like your shower. So it's worth spending a bit of money making this an enjoyable experience. Showers with good thermostatic controls start at around £200.

Tapmagic cartridges cost around £5 (www.tapmagic.co.uk). High quality taps with Ecotop cartridges installed are available from the Green Building Store from around £100 (www.greenbuildingstore.co.uk). The Green Building Store also sells an aerating shower head for around £30.

Further information

A free booklet on *Conserving Water in Buildings* is available from the **Environment Agency**. This includes sections on Showers and baths and Taps (www.environment-agency.gov.uk, 08708 506 506).

Water: Use less, save more (John Clift and Amanda Cuthbert, Green Books) describes 100 water-saving ideas.

The Water Book (Judith Thornton, CAT Publications) is a thorough guide to using water carefully. CAT also publishes low cost tipsheets on *Water Conservation* and *Making Use of Grey Water in the Garden* (www.cat.org.uk, 01654 705959).

Solar electric panels

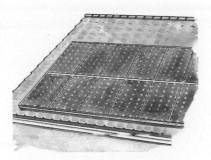

Solar electric panels turn light into electricity. They are commonly called 'photovoltaic' (PV) panels because they generate a voltage from the energy in light. As daylight is abundant, PV panels offer a truly universal opportunity for clean energy. Although they will only ever be part of the solution, given that the sun regularly sets, they are an important part: tried, tested and reliable.

Electricity from your roof

At the heart of a PV panel is a junction of two semiconductors, one positively charged and one negatively charged. The light that enters the panel bounces free electrons across the junction, creating an electric current. Using some smart electronics, this low voltage direct current is transformed into high voltage alternating current – the stuff we use every day in our homes.

Despite the complexity of the technology, PV panels are very simple to use. Once they are installed on your roof and plugged in, you can forget about them (usually). There are no noisy moving parts wearing out, no leaking pipes and no fuel or waste to lug around. Maintenance is minimal and the panels should last for decades (well beyond the 25 year warranty) as the solid state technology does not degrade.

Solar power and domestic needs

One of the few disadvantages of solar power is its poor match to our actual use of electricity at home. The sun is at its highest in the middle of the day when we need very little electricity, and absent altogether on winter evenings when the lights, television and kettle are all on. Happily this does not mean that PV panel owners have to go to bed at dusk – they simply switch to mains electricity instead. Most PV systems are fully integrated with the national grid and owners are paid for the surplus they export. You can store the power you generate in batteries, but this is a more complicated solution (page 34).

System components

Almost any roof or wall can generate electricity with a PV panel attached, but a pitched roof facing due south is best. South-east and south-west facing roofs are also suitable. You will get a decent output from a PV panel even under cloudy skies but direct sunlight pushes this output up three- or four-fold. Ideally you want a site free from shadow-casting obstructions, but don't be put off if there is a little shading. Your roof must of course be strong enough to support the panels.

PV technology comes in various forms. There are traditional panels which, like solar hot water panels (page 91), are bolted on to your roof and sit above a weatherproof rainscreen. Alternatively you can buy PV tile systems that replace your roof covering altogether, providing both a weather shield and electricity. These may be acceptable to planners even in conservation areas (ordinarily, PV panels do not need planning consent). Thin-film systems are also available, sold in large sheets which can be rolled onto your roof.

These are lightweight and lower cost but do not generate as much electricity as the other two technologies.

As well as the panels, a PV system will include an inverter which turns the low voltage DC output into 240V AC. The inverter is usually mounted on a wall inside your home, near the electricity consumer unit into which the PV supply is fed. Your supplier's package should also include installation, negotiation with the local power operator, commissioning and a very long warranty.

You can buy individual devices, such as fountains and garden lights, that are directly powered by small PV cells, but PV systems work just like the mains supply, powering all your lights and appliances across the house's wiring.

Power output and panel size

The average household in Britain burns around 4,500kWh of electricity every year. To generate this much, you will need a minimum of 30 square metres of PV panels. But don't settle for average: do what you can to get your electricity consumption down, by installing low energy lights and appliances, and you will need to generate far less. Reduce your bills with energy efficiency measures before you specify your PV panels.

In practice, most households with PV panels only generate part of the electricity they need. Your decision about how many panels to install is more likely to be determined by your budget and roof area. If you do end up generating more than you need, you will receive regular cheques from the power company.

The power output of a PV system is usually described as a peak output in kilowatts, such as 3 kWp. In practice you will actually generate about 85% of this peak in the best conditions. The total power output over the year (in kilowatt-hours) is a much more useful figure to work with, as you can compare it to the figures on your bills, so make sure your supplier quotes this. The figure should be based on actual performance in Britain.

Hidden costs

The manufacture of PV panels is very energy-intensive, largely due to the processing of the raw silicon, so for the first three years or so you will be recovering the energy required to make the panels. However you will then have four decades of fossil fuel-free electricity to look forward to. There are also environmental risks at the end of the panels' life due to the use of metals such as lead, cadmium and mercury in their manufacture.

Costs

PV technology is not cheap, but it does have a very long life. Full PV installations cost between £5,000 and £8,000 per kWp. A 3kWp system is likely to cost around £15,000 but will save you up to £250 per year in electricity bills. You may also be paid for any Renewable Obligation Certificates you receive (page 34) which could be worth around £100 per year for a 3kWp system.

Grants are available to support the installation of PV panels. Contact the Energy Saving Trust to find out what you are eligible for (www.energysavingtrust.org.uk, 0800 512 012). If you apply for a grant, you have to use an installer registered with the scheme.

Further information

PV installer details are available from the **Renewable Energy Association** (www.r-e-a.net)

For a complete guide, see *Practical Photovoltaics: Electricity from Solar Cells* (Richard Komp, available through Green Books).

Solar hot water panels

A solar hot water panel heats your water with the sun's rays. In the summer you can get almost all of your hot water from the sunshine that lands on your roof. You can then enjoy hot showers after sticky days in the knowledge that neither you nor the planet has to pay for heating the water.

Solar hot water panels are usually just called 'solar panels'. Confusingly, however, people also refer to panels that generate electricity as solar panels. These are better described as solar electric or photovoltaic panels (page 87).

Hot water from your roof

The simplest solar hot water panel is a glazed sheet of black metal mounted on the roof, facing the sun. A pipe winds across this piece of metal through which water flows, drawing heat from the panel. This pipe is connected to your hot water cylinder, which gets hotter as the water from the panel is pumped through it, before being directed back up to the panel in a continuous cycle. An electronic controller stops the pump when the temperature in the tank is high enough or when the temperature of the panel drops (e.g. at night).

Wherever you live, a solar hot water panel will produce at least half of your hot water over the year. You will still need your boiler for the darkest days of winter and for nights when all your hot water gets used up before the sun comes up, but your boiler will barely come on in the summer. In the winter months solar hot water panels will often preheat water for the boiler, but their relatively low output at this time of year means that they are rarely used to supply central heating. Hot water, however, is needed all year round.

Solar hot water panels produce the most heat when they face the sun. To get a decent annual output you therefore need a pitched, unshaded roof facing somewhere between south-west and south-east. On a flat roof you can either mount a panel on a pitched framework or install a panel specially designed to work on a horizontal surface.

Types of panel

There are lots of companies selling solar hot water panels and they will all tell you why their panel is the best. In practice, there is not much difference in their performance. The basic choice is between flat plate collectors and evacuated tubes. A flat plate collector is the design described above: a hot piece of metal with pipe attached. An evacuated tube collector is made up of a bank of glass tubes, each containing a long thin strip of metal which absorbs the sun's heat. This metal absorber is surrounded by a vacuum which reduces heat losses. An enclosed fluid draws heat from the absorber and transfers this to the main loop of water through a heat exchanger at the top of the tube.

Evacuated tubes are more efficient than flat plate collectors, but a slightly bigger flat plate collector will compensate for this. One square metre of flat plate collector per person is recommended with a minimum of four square metres.

Integration with your hot water supply

Plumbing a solar hot water panel into your existing hot water supply may or may not be straightforward. If you are not able to adapt your existing plumbing, you may need to fit a completely new hot water cylinder or an additional tank alongside it.

Most solar panels come as a package with limited options for integration with your existing plumbing. So always consider this aspect in detail when you are getting information and quotes from suppliers. A good supplier should take a keen interest in how your hot water supply is currently configured.

Direct systems

If you have a traditional hot water cylinder, the simplest way to integrate a solar hot water panel is to draw water from the cold pipe that feeds the cylinder, pump it through the panel, then put it back into the top of the cylinder. Your cylinder will then be fed with hot water from the top as well as cold water from the bottom. This is called a direct system because the water that goes through the panel is also the water that ends up coming out of your taps. It usually requires some form of 'drain back' plumbing to ensure that the panel empties out when the sun has gone, preventing the water freezing in the panel.

Indirect systems

In indirect systems, the fluid that goes through the panel is different from the water that comes out of your taps. This fluid, which normally includes anti-freeze, goes through a metal coil in the cylinder which transfers the heat to the water. As you are unlikely to have a spare coil in your cylinder to plumb the solar hot water panel in to, you will need to replace your cylinder to achieve this. Alternatively, if you have space, you can install a preheat tank with a coil in it for the solar panel. If you have a mains-pressure hot water supply, you must use an indirect system.

Combination boilers

If you have a combination boiler you will not have a hot water cylinder at all, which makes integration with a solar hot water panel very difficult as the panel has to dump its heat somewhere. You can use a preheat tank with some of the more advanced combination boilers but most are designed for use with mains cold water only. Your only option then is to wait until you need to replace the boiler before installing a solar hot water panel.

Central heating too?

If you have a well insulated, airtight house, underfloor heating and room for a large hot water cylinder – a 'thermal store' – you could potentially upgrade to a solar hot water system that provides some heating in spring and autumn as well as hot water. Get advice first to see if this option is appropriate for your circumstances.

Installation

Solar hot water panels are relatively easy to install as they are designed to sit above the existing roof, rather than replacing it. The main technical difficulty is securing the panel and ensuring that the penetrations made for the fixings and pipes do not leak when it rains. A good installer will provide you with a guarantee for every aspect of the system including works to the roof.

Costs

A fully installed solar hot water panel with appropriate connections to your hot water supply should cost you between £2,000 and £4,000. Unfortunately the solar industry has its fair share of sharks, so beware of door-to-door salesmen offering you amazing deals.

You can buy a solar panel kit from B&Q, including a flat plate collector and preheat tank, for £1,800. For an even cheaper solution, make the panel yourself: the Low Impact Living Initiative runs courses on DIY solar water heating (www.lili.org.uk).

Grants are available to support the installation of solar hot water panels. Contact the Energy Saving Trust to find out what you are eligible for (www.energysavingtrust.org.uk, 0800 512 012). There are many local grants schemes as well as a national scheme. If you apply for a grant, you have to use an installer registered with the scheme.

Further information

The **Solar Trade Association** is a useful place to begin looking for installers in your area (www.greenenergy.org.uk/sta).

A free information sheet on *Solar Water Heating* is available from the **Centre for Alternative Technology** (www.cat.org.uk, 01654 705989). CAT also publishes two books on solar hot water: *Tapping the Sun: a guide to choosing a solar hot water system* and *Solar Water Heating: a DIY guide* (01654 705959).

For more information on using solar energy for room heating using a thermal store, see the website of **Greenshop Solar** (www.greenshopsolar.co.uk).

Televisions & electronic goods

We now burn more electricity sitting in front of the TV, computer and hi-fi than we do cooking, washing or keeping things cold. The rise of red-eyed blinking boxes has wiped out many of the energy savings from the improved energy efficiency of 'white goods' such as fridges and washing machines. So if you want to green up your energy consumption, don't ignore these important parts of your daily electricity use.

The hidden costs of manufacture

The resources and fuel that go into making a desktop computer weigh as much as a rhinoceros. Electronic goods are designed to be sleek, but their discreet design hides huge environmental costs during manufacture. Unfortunately many electronic products have short lives because new models with enticing features are always just around the corner. You should therefore do what you can to slow down this rapid turnover of new products.

The easiest way to achieve this is to buy second-hand. If you are determined to buy the latest model, make sure your old kit goes to a good home either by selling it on or donating it to one of the many charities that organise the reuse of computers and other electronic equipment. By law all electrical and electronic equipment must now be recycled, so check with retailers and your council to see how you can get rid of defunct equipment.

Prioritise durability and adaptability: assess how equipment will cope with abuse, and consider future changes in its use. Make sure a new television is 'HD ready' (ready for high definition broadcasts) and a new computer has space for memory upgrades.

Screen technology

Old-fashioned cathode ray tube televisions have almost disappeared from the market (though not the second-hand market) despite the fact that the technology is more energy-efficient than either liquid crystal display (LCD) or plasma screens. There is now little difference between the energy performance of LCD and plasma but as plasma screens tend to be used for really big televisions (40 inches or more), they burn lots of energy thanks to their sheer size. Your best option for a new flatscreen television is therefore a moderately sized LCD screen with very low stand-by power consumption.

When it is turned on, the electricity your television requires depends on how high you set the brightness of the screen. As televisions often leave the factory with the screen brightness set very high, try adjusting your set. You may find you get a better picture at a lower brightness.

Set-top boxes

Set-top boxes decode digital signals for ordinary televisions. They tend to be left on and so add to the background energy we burn in our homes. If, however, you buy an integrated digital television (IDTV), you do not need a separate set-top box to watch terrestrial digital channels (Freeview). If you are buying a new television, look for the blue 'Energy Saving Recommended' label as this is only given to IDTVs with a stand-by power consumption of less than 1W.

If you are happy with your television but want to upgrade to digital, you will have to buy a separate set-top box. These are now included in the 'Energy Saving Recommended' labelling scheme, so look for the label in the showroom or check the Energy Saving Trust's product database (www.energysavingtrust.co.uk). The label is also given to digital video recorders. Like IDTVs, these are available with the set-top box digital tuner built in. Always avoid buying two boxes when one will do.

Unfortunately set-top boxes for satellite television need more energy than ordinary decoders (10W or more). As these boxes come as part of a satellite subscription package, you don't have much choice here – other than sticking to terrestrial digital television.

Looking for low energy

Finding information about the energy consumption of electronic goods is not always easy. Some televisions do have energy labels, but most do not. As your average shop assistant is unlikely to be of any help, turn to the internet instead. Go to manufacturers' websites, select specific products and look through their specifications. Power consumption and stand-by are usually there, somewhere near the bottom of the list. This approach can be time-consuming and frustrating, as not all brands make this information available, but it is informative. It is relatively straightforward for televisions and computers, harder for audio equipment and games consoles (Playstation 3 burns a whopping 380W).

Happily it is not difficult to identify the electronic products with the very best energy performance. Simply check the recommendations at www.topten.ch. This website is not in English, but don't be put off as it is straightforward to navigate.

The only way to know for sure how much power any electronic product is using is to buy a portable meter which you plug into the wall socket. You then plug the device into the meter and a display tells you how much power is being drawn – a great way to learn about the relative impact of all the electronic gizmos in your home.

Stand-by control

The familiar green mantra to turn your television and computer off rather than leaving them on stand-by is often misunderstood. It may not be enough to get up, press the power button and watch the little red light fade. To make a difference to stand-by power you usually have to flick the switch at the power socket on the wall, which can be inconvenient or worse. If you have equipment that you have to leave on, such as video recorders, pay special attention to buying products with the lowest possible stand-by power consumption. If you simply cannot reach the wall socket, buy a remotely operated switch.

Costs

The energy performance of electronic goods is primarily a function of their features and size, and rarely has an independent effect on price. You will however have to pay more for high quality, robust hardware that is designed to last.

The Plug-in Mains Power and Energy Monitor costs £25 and the Bye Bye Stand-by remotely operated stand-by switch costs £20. Available from The Green Shop (www.greenshop.co.uk, 01452 770629) or www.ecoelectricals.co.uk.

Further information

The *Greenpeace Guide to Greener Electronics* assesses the efforts of computer and mobile phone companies to reduce their environmental impacts (www.greenpeace.org.uk). *Ethical Consumer magazine* rates companies against a wider set of environmental and ethical criteria (www.ethicalconsumer.org).

Toilets

Two things go down the majority of the nation's toilets: drinking water and our very own organic waste. The best way to green up your toilet use is to reduce the amount of drinking water you flush away. You may also want to explore more ecological ways of dealing with the organic waste, but the composting toilet, that most evocative symbol of green living, is a rare choice among those who already have a connection to a mains sewer.

Don't put a brick in it

You can reduce the amount of water flushed down a toilet by displacing some of the water in the cistern. Water companies offer free 'hippos' for this purpose, but all you need is a filled and sealed plastic bottle. Don't use a brick as this will disintegrate and damage or block your plumbing. Above all, don't reduce the flush so much that the toilet doesn't do the job properly and you end up flushing twice. Experiment with different sizes of bottle to see how far you can go without suffering this problem.

Traditional toilets with a lever on the front can be retrofitted with a device that stops the flush as soon as you let go of the lever, so you can control how much water is released for each flush.

Low-flush toilets

If you are replacing your toilet or fitting out a new bathroom, install the most water-efficient toilet available. The maximum flush allowed by law is six litres, but you can do better than this. A well-designed dual-flush toilet needs only four litres to do the maximum job and two litres for the lesser task.

Most dual-flush toilets have buttons which operate valves that release different volumes of water down the pan. However you can also buy a dual-flush toilet with a traditional lever-operated siphon mechanism. The smaller flush is achieved by pressing and holding the lever. The advantage of the buttons and valve system is that any user can see the choices available to them. The advantage of the lever and siphon system is that it cannot leak. If you have a valve toilet you must keep an eye on it to make sure the valves do not get damaged, as this can lead to a slow leak of water into the pan, resulting in substantial water loss over time.

Rainwater and grey water

You can radically reduce the amount of drinking water you flush down the toilet by replacing it with rainwater. Toilets are the primary beneficiaries of rainwater collection systems because water purity is not critical (page 71). It might seem like a good idea to use the waste water from showers and baths – grey water – to flush your toilet but unfortunately this water is full of bacteria which rapidly multiply when the water is stored and so starts to stink in no time. Grey water is best reused immediately on your shrubbery, if you can get it there, or not at all.

Drainage and composting

If you are not connected to a sewer there are many ways of dealing with the outflow from your toilet. The most common are cesspools and septic tanks: the former is a sealed container that is emptied regularly by a sludge lorry, the latter percolates the liquids into the surrounding ground and so needs emptying much less often. If you want to overcome dependence on the lorry and make better use of the material being lost to the sewage plant, a compost toilet is a better option.

Compost toilets come in many forms. The simplest rely on a box (or two boxes, one for each type of waste) immediately under the seat. More sophisticated models feed out-of-sight basement digesters. If you have space, you could even plant a reed bed which will turn your waste products into fast-growing reeds and clean water. These options require careful assessment, not least because of the health and safety issues involved. If you are interested, consult the books listed below and seek professional advice.

Costs

The Interflush water-saving device for traditional lever flush toilets costs around £20 and can be bought direct from Varyflush Ltd (www.interflush.co.uk, 0845 045 0276).

Six-litre dual-flush toilets can be found in any bathroom shop in all price brackets. For a high quality four-litre dual-flush toilet, consult specialist suppliers: the Ifö range is available from the Green Building Store (www.greenbuildingstore.co.uk, 01484 461705), starting at around £300.

Further information

A free booklet on *Conserving Water in Buildings* is available from the **Environment Agency**. This includes sections on water-efficient WCs, and waterless and vacuum toilets (www.environment-agency.gov.uk, 08708 506 506).

Free information sheets on *Composting Toilets* and *Small-scale Sewage Treatment* are available from the **Centre for Alternative Technology** (www.cat.org.uk, 01654 705989). For more detail, see

Lifting the Lid: An ecological approach to toilet systems (Peter Harper and Louise Halestrap) and *Sewage Solutions: Answering the call of nature* (Nick Grant, Mark Moodie and Chris Weedon), both from CAT Publications.

The **Low Impact Living Initiative** (www.lowimpact.org.uk) runs a regular course on sustainable water and sewage.

Underfloor heating

Despite the best efforts of designers to produce attractive radiators, our homes undoubtedly look better without them. The popularity of underfloor heating is primarily driven by our desire to liberate our walls from these imposing lumps of metal. By turning the entire floor into one big radiator, underfloor heating puts the heating system in its place: in touch but out of sight.

Water vs. electricity

Most underfloor heating systems use hot water supplied by a boiler or other heat source. The water is pumped around all the floors of your house in exactly the same way that it is pumped to radiators.

Some underfloor heating systems use electricity to provide the heat instead. These are easier to install than wet systems, but are not at all eco-friendly. Electricity is the dirtiest and most inefficient domestic fuel because so much energy is lost in the power station (page 32). It should only be used for heating as a last resort. Electrical systems are also much more expensive to run.

Efficient and effective

Because radiators are relatively small, the temperature of the water that flows through them has to be kept quite high in order that enough heat can be released into the rooms of your house to keep them warm. As underfloor heating uses a much larger surface area to emit heat, it can run at a lower temperature. As modern condensing boilers run more efficiently at lower temperatures, this difference makes underfloor heating a greener choice. But only just: do not rip out your radiators and install underfloor heating for this reason only.

Underfloor heating is also more effective in delivering heat to where it is needed. Radiators create currents of air that take the heat to the top of a room, where we need it least, and create draughts of cold air at floor level. Underfloor heating is more comfortable because it radiates heat gently across the whole room. There are no hot and cold spots and no currents of dry, dusty air. Rooms with warm surfaces and stable air do not require the high air temperatures of radiator-supplied rooms: you may be able to turn down your thermostat, save energy and still be more comfortable.

Not for everyone

Small lumps of metal heat up quickly; concrete or timber floors do not. The big disadvantage of underfloor heating is its sluggishness. If your home is cold and you want to heat it up rapidly, underfloor heating will not do the job. Underfloor heating works best with well-insulated, energy-efficient homes that do not lose heat quickly and stay at a fairly stable temperature across the day. It is therefore most suited to constantly occupied homes.

Underfloor heating can also exacerbate overheating problems. If you have rooms with a lot of glazing that heat up quickly when the sun comes out, you will be in trouble if your floors continue to emit heat long after the heating system has been switched off. Such problems can be avoided by reducing the risk of overheating with good shading (page 21).

Floor finishes

Underfloor heating can be used with any floor finish, but solid floors of stone or tile are best. Underfloor heating can be laid beneath timber floors as long as the timber is dry. Existing floorboards in centrally heated houses will be suitable, but new timber floors should be given time to acclimatise to the building first. Avoid fitted carpets as these act as insulation and slow down the heat flow to the room. On ground floors insulation must be laid first or you will simply heat the ground.

Controls

Underfloor heating is laid as a series of pipes, one for each floor or room. The pipes from each room go back to a set of control valves called a manifold. Thermostats in each room are electronically linked to these valves so that whenever the temperature in a room reaches the desired level, set on the individual thermostat, the relevant valve closes and the flow of hot water through the floor is stopped. The manifolds are in turn connected to the output of your boiler or alternative heat source.

Costs

As underfloor heating is more efficient than a traditional radiator system, it works out cheaper to run, especially with a condensing boiler. However it is likely to be more expensive to install if your existing floors have to be taken up or even replaced.

Further information

A free information sheet on *Underfloor Heating* is available from the **Centre for Alternative Technology** (www.cat.org.uk, 01654 705989).

Ventilation

Good ventilation is essential for health, life and quality of life. Poorly ventilated homes can be stuffy, damp and dangerous, especially if fires or boilers do not have a ready supply of oxygen, so it is essential that any improvements you make to insulation and draught-proofing do not leave you with poor air quality. You must consider the ventilation of your home at the same time.

Getting in control

Most houses in Britain have far too much ventilation, otherwise known as draughts. Even if you can't feel a draught in the winter, your home is probably leaking warm air through every little crack and junction. Your aim should be to reduce this uncontrolled air infiltration as best you can and replace it, when necessary, with ventilation that you can control. In doing so you will reduce the energy you waste but maintain or improve your air quality. In the summer, when there may be too much heat in your home, you can open the windows.

Even when you have blocked up your fireplaces and done your very best to draught-strip your floors, windows, doors and letter box (page 27), there may still be enough fresh air penetrating the building to provide decent air quality. You will however need some form of ventilation in kitchens and bathrooms where water vapour and other fumes build up.

Extractor fans

If you have extractors in your kitchen and bathroom, consider whether they are doing a good job – you could save a lot of energy with an upgrade. If your extractor is operated by your light switch, stays on for too long, makes a lot of noise and lets too much air in when it's turned off, you have scope for improvement.

Modern extractor fans can be triggered by humidity sensors so they operate only when they really need to. They have baffles on the back to prevent unwanted air flow when they are not in use. The best are quiet and low powered (10W). Overall they can cut electricity consumption by up to 80%.

Heat recovery room ventilators

Extractor fans throw warm air away. This energy loss can be radically reduced with heat recovery room ventilators. These look like extractor fans but they bring air into the house as well as exhausting air to the outside. The two streams of air cross through a heat exchanger which takes 80% of the heat from the outgoing air and puts it into the incoming air, so you get a steady stream of fresh, warm air and very low energy losses. Some heat recovery room ventilators have very low fan power (2W) so they can be left running continuously, helping to lower humidity, dry clothes and reduce dust mite populations which thrive in warm, humid conditions. They are relatively easy to install and can directly replace wall-mounted extractor fans.

Passive stack ventilation

Another way of ventilating kitchens and bathrooms without using any electricity is 'passive stack ventilation': warm, moist air from these rooms is encouraged to naturally rise through special ducts to a vent in the roof. But unless you have a perfectly positioned unused chimney, retrofitting the ducts can be very difficult.

Fresh air when you need it

Many modern houses have 'trickle vents' in their windows and doors: narrow grilles which can be opened to improve background ventilation. Unfortunately these tend to leak lots of air and heat even when they are closed, so they generally do more harm than good. A better way of providing background ventilation, common in u-PVC windows, is secure partial opening of windows. This lets air in when needed but does not compromise the airtightness of the window when it is closed.

If you are at all concerned about the air quality in your home, open the windows fully. If you do this briefly, for a few minutes, you will flush out stale and moist air, replacing it with fresh air, with relatively little heat loss. This is because most of the warmth in your home is held within the building itself – your floors, walls and ceilings – and not in the air, which has a low heat capacity. So although the steady drain of heat through draughts adds up to significant heat loss, a quick refresh will not make much difference.

Whole house heat recovery ventilation

The most advanced form of eco-ventilation is a whole house system that delivers fresh air to the living spaces, removes stale air from kitchens and bathrooms and transfers the heat in the outgoing air to the cool incoming air. These systems can be driven by fans, requiring some electrical energy, or by wind cowls on rooftop air inlets which drive fresh air down into the house.

Such systems are rarely suitable for retrofitting, not least because of all the ducts that have to be hidden in the ceilings and walls. More importantly, this level of ventilation is only necessary when the uncontrolled air flow through the fabric of the house is negligible. However enthusiastic your draught-stripping may be, you are unlikely to achieve this. You might nonetheless consider this option if you suffer from asthma, as this level of ventilation, combined with a thorough clean of your bedding and fabrics, will keep humidity and dust mites to a minimum.

Reduce the need

You can reduce the need for ventilation by reducing the moisture, fumes and toxins you release in your home, for example by turning the shower on only at the very beginning and end of your wash (page 84), keeping lids on pans (page 20) and doing away with inefficient gas fires. Modern boilers do not draw oxygen from the room they are in but through a balanced flue which brings in air as well as exhausting it (page 8). Reduce air pollutants with eco paints (page 63) and avoid smoking indoors.

Costs

Simple energy-inefficient extractor fans can be bought for £20. You will have to pay three or four times this for one with humidistat control and low power consumption. The Vent-Axia HR25 low-powered heat-recovery room ventilator costs around £190.

The components of a whole-house heat-recovery ventilation system will cost between £1,000 and £2,000 depending on the size of the house. The installation costs for a retrofit may be considerable if ductwork has to be hidden in ceilings or boxed in.

Further information

The **Energy Saving Trust**'s free online library of best practice guides includes *Energy efficient ventilation in dwellings* and *Energy efficient refurbishment of existing housing* (www.energysavingtrust.org.uk).

Walls

The walls of your home do many different jobs at once. They support the roof and the floors, keep the heat in and keep the rain, cold and noise out. You will have to act fast if the roof is falling down, but millions of people survive with poorly insulated walls simply by turning the heating up. Getting your walls to work properly in every respect is therefore a top priority for greening up your home.

Get cosy

Walls with no insulation throw heat away with abandon. Installing insulation can be a major task, given the area you have to cover, but you will definitely feel the results: warm walls make for a green and cosy house. Remember that it is only your exposed external walls that you have to fix, as you can assume that walls adjoining other properties will gain as much heat as they lose.

As well as cutting your heating bills by up to a third, insulated walls improve the comfort of your home, prevent condensation forming on your walls and keep the heat out on hot summer days. If however there are any existing problems with your walls, such as damp, you should address these first or you may make matters worse.

Cavity wall insulation

Many houses built in the twentieth century have double-skin walls with a cavity between the two layers. The aim of the cavity is to prevent moisture getting through the wall and into the house. It is also an ideal place to retrofit insulation by blowing it in through holes drilled in the outer wall. The insulation is typically made of polystyrene beads or fragments of mineral fibre which fill up all the nooks and crannies inside the cavity but prevent moisture from tracking across to the inner wall.

Cavity wall insulation requires professional installation, but this is usually straightforward and takes less than a day. Once the insulation has been blown in to the walls, the holes are made good and the building is left with no significant changes to its appearance.

Solid wall insulation

Most houses built before 1930 have solid walls with no cavity. You can only improve these walls by adding insulation to the inside or to the outside. Either way, you are likely to need professional help for anything but the most basic interior insulation. Covering entire walls with a new layer of insulation and a new finish is a big and messy business, so it is worth undertaking with other renovation tasks.

The simplest way of adding insulation to a solid wall is to stick it in on the interior face. You can buy plasterboard which is designed specifically for this purpose with a layer of high performance insulation already attached. The thicker the insulation layer, the more energy you will conserve. You can do better still by building a new wall in front of the existing one with timber uprights (studs). You can then pack the gaps between the studs with 100mm of insulation or much more. Internal insulation must include a barrier (a polythene sheet under the plasterboard) to stop water vapour getting into the wall where it could condense on the masonry behind the insulation. This barrier is included in insulated plasterboard products.

The depth you choose will partly depend on how much of your room you are willing to lose (only the exterior walls need insulating, remember) and how many window sills and architectural details you have to work around. If you do not want to lose any of your room space, you could still install thermal wallpaper. This thick lining paper will not do much to cut your heat losses but it will give you warm, condensation-free walls.

Exterior insulation is much more effective than interior insulation because the whole house is wrapped in a new warm blanket. It is however a major task as you must create a new external weathershield for your house, under which you can pack lots of insulation. This requires careful design and installation, as it must be very durable, but it can improve the look of your house as well as its thermal performance. If you have an old house with lovely cornices *and* an attractive façade, you can always leave the front and insulate the back. Many Victorian houses have far more exposed wall area on back extensions than at the front.

New walls

The choice of building materials for new exterior walls is extensive. Deep green options include straw, earth, cob (a mixture of earth and straw), hemp and even old car tyres. If any of these options appeal, you must seek advice as they all require specialist knowledge and skills to get right. Another natural material, timber, is more widely used in new construction and is a practical green alternative to concrete blocks. However you must still find builders with the right experience as a timber wall works in a completely different way to a brick and block wall. An advantage of timber in tight spaces is that the wall does not need to be as thick as a block wall to achieve the same heat-retaining properties.

Most walls are still made of two layers, so whatever the main structure is made out of, you can still match the exterior to the existing finish of your house if necessary. If you have a choice, the best options are again natural, unfired materials such as stone and naturally durable wood (i.e. woods which do not require preservatives).

There are green options even with bricks and concrete blocks. Reclaimed bricks are an excellent choice and will be reusable again in the future if you use flexible lime mortar rather than cement to stick them together. Lightweight concrete blocks made from pulverised fuel ash make good use of a waste material from power stations, but are still very energy-intensive to manufacture.

Costs

Cavity wall insulation typically costs around £250-£300 for an average home. Grants are often available – contact the Energy Saving Trust for information (see below).

Insulation-backed plasterboard costs around £19 per square metre for the thickest insulation (60mm). Sempatap thermal wallpaper is similarly priced, though local discount schemes are available (www.mgcltd.co.uk, 020 8337 0731).

The cost of external insulation is high because a new weather-proof exterior has to be added to the building. Depending on the architectural complexity of the outside of your house, you may need to hire an architect before you get near seeking quotes from builders. You may also need planning permission, especially if you live in a conservation area.

Further information

Talk to an energy efficiency advisor before you install wall insulation. Call the **Energy Saving Trust**'s national helpline on 0800 512 012 to be put in touch with a local advisor. The Trust's website also offers useful information on wall insulation (www.energysavingtrust.org.uk) and its free online library of best practice guides includes *Domestic energy primer: an introduction to energy efficiency in existing homes, Cavity wall insulation: unlocking the potential in existing dwellings, Internal wall insulation in existing housing, External insulation for dwellings* and *Advanced insulation in housing refurbishment.*

To find an insulation installer, contact the **National Insulation Association** (www.nationalinsulationassociation.org.uk, 01525 383313). Cavity wall insulation installers should be registered with the **Cavity Insulation Guarantee Agenc**y (www.ciga.co.uk).

For information on natural building materials, see *Natural Building: A Guide to Materials and Techniques* (Tom Woolley, Crowood Press) or *The Whole House Book: Ecological Building Design and Materials* (Cindy Harris and Pat Borer, CAT Publications).

For up-to-date information about building materials and the range of products available, see the **GreenSpec** website (www.greenspec.co.uk).

To find builders experienced in using unusual building materials, see the listings in the **Green Building Bible** (Green Building Press, www.greenbuildingbible.co.uk) or search the membership directories of the **AECB** (www.aecb.net) and **Green Register** (www.greenregister.org).

Washing machines

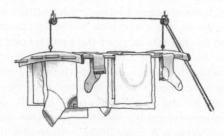

The domestic appliance that has done the most to reduce the drudgery of daily housework is undoubtedly the washing machine. Although the greenest way to wash your clothes is with sink, soap and a hand-operated mangle, this is really only viable for those of us with a basketful of hair shirts. The washing machine is here to stay, so make sure it does the best possible job for you and for the environment.

Buy cheap, buy twice

These days you can buy a washing machine with an excellent energy label for under £200. But it won't be a good green choice.

Washing machines are victims of our collective determination to prioritise price over quality. With few exceptions washing machines are no longer built to last 15-20 years, and basic repairs to some models can cost almost as much as buying a new machine. As a result, we think nothing of replacing a washing machine every five to seven years, clocking up huge hidden environmental costs because of all the energy and resources that go into making new products. So if you are in the market for a washing machine, think long-term and invest in a high quality, durable machine.

The energy label

Having opted for quality, pay attention to the energy label. Washing machines are described with three letters, such as 'AAA'. The first letter in this threesome is the important one as it describes the energy efficiency of the model. Look for an A+ or A++ machine for the lowest energy costs and carbon emissions. The second letter describes washing performance (how clean the clothes come out) and the third letter describes spin drying performance (how dry your clothes come out). If you must have an energy-guzzling tumble dryer, make sure your washing machine's spin cycle is A-rated in order to reduce the time your clothes spend in the dryer.

The label also tells you how much water is used per wash. Although modern washing machines use far less water than older models they are still thirsty beasts, so make water efficiency one of your criteria for comparing machines. Avoid top-fill machines as they need lots of water to get everything wet.

Hot and cold fill

In the past many washing machines drew hot water from the household hot water cylinder to wash the clothes. This made sense because most people heat their water with gas or oil boilers which are much more energy-efficient than direct electric heating. However most washing machines are now 'cold fill', drawing cold water and heating up what they need using electricity. The manufacturers claim that this is the most energy-efficient approach because a) modern machines need very little hot water and b) if you draw hot water from your cylinder you end up wasting heat from the water that gets left in the supply pipes once the machine has taken what it needs.

Don't be surprised, therefore, if even the most environmentally conscious brands no longer offer hot fill machines.

Wash with care

There's little point in buying a high quality, durable, energy-efficient and water-efficient washing machine if you then put every shirt through a boil wash as soon as it's off your back. The following are some basic tips for an environmentally friendly approach to clothes washing:

- Take care to separate clothes that need a wash from those that can be worn again.
- Airing or cold-soaking clothes will remove light odours without the need for a machine wash.
- Select low temperature washes where possible.
- Always run full loads, as two half loads use much more energy than one full load.

Drying clothes

A tumble dryer uses more than twice the energy of a washing machine. Fortunately this is one appliance that is still owned by a minority of households in the UK, so do your bit to keep it that way. There is no such thing as a green tumble dryer.

Not everyone has a garden to hang a washing line but we all have spare space on our ceilings. A ceiling mounted clothes airer that can be raised to take advantage of the warm air at the top of a room is a very effective way of drying clothes. You just need to make sure (as with a tumble dryer) that the room is adequately ventilated to allow moisture to escape.

Costs

A top of the range washing machine will cost you between £500 and £1,000. The more expensive brands are also more expensive to repair but they should break down far less often than the cheap models and last much longer.

Ceiling mounted clothes airers start at £65 from the Natural Collection (www.naturalcollection.com, 0845 3677001).

Further information

The **Energy Saving Trust**'s database of Energy Saving Recommended Products includes lots of washing machines (www.energysavingtrust.org.uk, 0800 512 012). It also includes tumble dryers, so if you must have one, look here first. In shops, the products included in this database will display the blue 'Energy Saving Recommended' label as well as the standard energy label.

A detailed ranking of washing machines based on their water and energy efficiency is provided by **Waterwise** (www.waterwise.org.uk).

A useful guide to buying, using and maintaining washing machines can be found at www.washerhelp.co.uk.

Water turbines

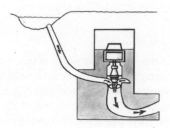

If you live next to a tumbling stream, you have an unbeatable energy resource on your doorstep. Unlike solar and wind power, micro-hydro generators never stop. Unlike wood burning, the fuel is free and leaves no waste to clear up. There are no carbon emissions to worry about, and you even escape the localised environmental damage that large hydro projects create. If only we could all have one.

Mill wheels and water turbines

Historically, water power was harnessed in Britain with big slow wheels that turned heavy mill stones. Water turbines, however, prefer a rapid spin to a slow push so although it is possible to convert waterwheels to produce power, you will get far better results with a system that is designed for electricity generation from the outset.

The two critical variables that affect the power output of a water turbine are the flow of water through it and the height that it falls before hitting the turbine. The greater the 'head height', the more the water will speed up under gravity – if you put a turbine directly into a river without creating any head height to accelerate the water, you will only get a trickle of energy from it. A slow, meandering stream is therefore not much use. If your land drops by a few metres (one at the very least), a water turbine may be worth investigating.

Assessing head height and flow

If you are not sure how much your land drops, you can measure the drop with a hosepipe. Get someone to hold one end of the pipe at the point where the stream enters your land. Take the other end as far as you can down the stream. Pour water into the pipe from the top and raise the other end until you reach the point where water no longer spills out – i.e. the same height as the water going in at the top (you may need a step ladder if the land is steep). You can then measure this height to the ground with a tape measure and repeat further down the stream if necessary.

Flow is much harder to measure. If you think you have enough head height, and your stream is more than a trickle, get a site assessment done which will include a measurement of the flow and an estimate of its variability over the year. You will then have enough information to work out how much annual electricity you will get from a water turbine. If your stream can drive a 1kW turbine, you should generate more than enough electricity to meet your needs.

System design

To supply a water turbine with fast flowing water, you need to channel the water out of the stream, filter it for leaves and other large contaminants, accelerate it down a pipe, let it flow through the turbine, then return it to the stream. There are different turbine designs for different levels of water throughput, but they are all at least 50% efficient.

Water turbines are often installed by people who live off the electricity grid and rely on batteries to store the electricity they generate. If you have a grid connection, a water turbine can be integrated with your electricity supply so that you export surplus power and buy it back when the turbine is generating less power than you need (page 34).

Getting permission

If the best place to tap the water lies outside your land, or the stream is itself a property boundary, you will need to get the permission of your neighbours. You also need to get approval from the Environment Agency to extract water from a water course. Local planners may also be interested if there is any substantial housing for the system above ground.

Costs

Water turbines are expensive so it is worth paying for a site survey first and then seeking quotes for supply and installation. The actual costs of installation will depend on the extent of the ground works and building required to divert and channel the water. These costs can be substantial so a full project cost of £20,000 is not unusual, much more for larger turbines. Fortunately water turbines have long lives and low maintenance costs.

Grants are available to support the installation of micro water turbines. Contact the Energy Saving Trust to find out what you are eligible for (www.energysavingtrust.org.uk, 0800 512 012). If you apply for a grant, you have to use an installer registered with the scheme.

Further information

A free information sheet on micro-hydro systems is available from the **Centre for Alternative Technology** (www.cat.org.uk, 01654 705989). For more detailed information, see *Going with the Flow: Small scale water power* (Billy Langley and Dan Curtis, CAT Publications, 01654 705959).

For information on hydro-electric power in general, including a guide to 'mini hydro', see the **British Hydropower Association**'s website (www.british-hydro.org).

A clear introduction to domestic water turbines is provided online by the company **Hydrogeneration** (www.hydrogeneration.co.uk).

Pedley Wheel is a useful source of information on the reuse of waterwheels (www.pedleywheel.com).

Wind turbines

The British Isles are famously wet and windy. Continental Europe shelters behind the British and Irish bulwark, over which North Atlantic weather systems relentlessly break. Yet despite this plentiful wind resource, wind turbines are still a relatively uncommon sight in these islands. This is changing, thanks to the unbeatable low-carbon credentials of modern wind turbines, but do not assume that a wind turbine is a quick and easy way to green up your home. Your personal wind resource may simply not be good enough.

The right kind of wind

Wind turbines need a good clear view of a steady flow of wind. Any obstruction to this view will block the wind and create turbulence, sending wind currents all over the place. In these conditions, a wind turbine will constantly chase the wind and never get up to speed. It will also wear out very quickly.

The power output of a wind turbine is proportional to the cube of the wind speed: when the wind speed doubles, the power output goes up eight-fold but when the wind speed halves, the power output collapses. So a good wind will deliver great results but anything else will be almost useless. Turbulent wind produces very little power.

Local wind speeds

You can get a basic measure of the average unobstructed wind speed where you live by consulting the online wind speed database of the British Wind Energy Association (www.bwea.com/noabl). Different values are quoted for different heights above ground. If you can't reach an average wind speed of at least five metres per second, the viability of installing a wind turbine is doubtful (remember that any obstructions will radically reduce the quoted values). You can get a more accurate measure by setting up an anemometer and recording the wind speed over a period of time. This is recommended if you are thinking of investing serious money in a wind turbine.

Assessing your options

If you live in a built-up area or your home is close to other buildings or trees, think twice. Even when there is a strong wind blowing, the local topography will ruin it for you. If however you have a clear view of the prevailing wind, a turbine is definitely worth considering.

You can mount small wind turbines on a rooftop gable end, but get advice first to ensure that the structure is strong enough. You will get far better results if you install a wind turbine away from your house on its own mast, high above any obstructions. If you are lucky enough to have wind worth tapping, this is how to make the most of it. The mast-mounted strategy also takes the noise of the turbine away from your windows and the vibrations away from your walls. Either way, you are likely to need planning consent.

Power output

All wind turbines have a rated power output. This is a measure of how much power the turbine will produce in standard conditions

– usually a wind speed of 12.5 metres/second. A typical roof-mounted wind turbine will give you 1kW at 12.5 metres/second (on the rare occasions when this wind speed is reached) whereas a typical mast-mounted turbine will give you five times this output or more.

Ask your potential supplier for a copy of the power curve of the turbine. This will show you how the power varies with the wind speed, including the speed at which the turbine cuts in and the speed at which it cuts out. If the wind is blowing too hard, turbines shut down for safety reasons.

Be wary of optimistic claims about how much electricity a wind turbine will provide over the course of a year. This will always be very dependent on your local circumstances. Ask for a range of estimates based on different average wind speeds.

On or off the grid?

Many wind turbines are installed by households that have no connection to the electricity grid, usually in conjunction with large batteries. For those of us who enjoy a grid connection it makes more sense to export any of the surplus electricity generated by the wind turbine to the grid rather than storing it in bulky batteries (page 34).

Grid connection is arguably an overly complex and expensive strategy for very small wind turbines, given their low power output. Alternatives currently offered by turbine designers include using the power as and when it is generated or using it to heat water which can be stored much more easily than electricity.

Costs

The range of prices for wind turbines reflects the range of the technology. Domestic models start at under £1,000, but a typical 1kW grid-connected system will cost you at least £3,000 plus installation costs.

You should seek quotes that include a full site survey, installation and commissioning. You should also negotiate some form of maintenance contract as wind turbines need looking after.

Grants are available to support the installation of domestic wind turbines. Contact the Energy Saving Trust to find out what you are eligible for (www.energysavingtrust.org.uk, 0800 512 012). If you apply for a grant, you have to use an installer registered with the scheme.

Further information

Details of wind turbine suppliers can be found on the **British Wind Energy Association** website (www.bwea.com).

A free information sheet on *Domestic Wind Power* is available from the **Centre for Alternative Technology** (www.cat.org.uk, 01654 705989). For more detailed information, see *It's a breeze: A guide to choosing windpower* (Hugh Piggott, CAT Publications, 01654 705959). Hugh Piggott's website is also a great source of insight into small scale wind power (www.scoraigwind.com).

Windows

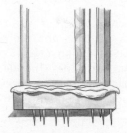

Windows do a great job admitting daylight to our homes but they also throw heat away with abandon. The daylight keeps our electric lights turned off but the heat loss keeps the boiler turned on. Happily this eco-conflict can be resolved, or at least minimised, by improving the heat-retaining properties of windows without any loss of light.

Window woes

On a winter's day when it is freezing outside and your indoor air temperature is 20°C, the inside surface of a single glazed window will be a mere 5°C. When warm air meets this surface it immediately cools and falls, creating a chilly draught in the room. Water vapour in the air condenses on the surface and trickles down. The window itself radiates cold on to anyone who stands in front of it. In all these ways, cold windows create cold, damp and uncomfortable rooms. The more you improve your windows, the fewer of these problems you will face.

How windows work

Letting in light is easy – all you need is glass – but stopping the heat escaping through a window is a much more difficult task.

Heat is lost through conduction, convection, radiation and air infiltration. Each of these can be tackled through improvements in window design.

Heat is directly conducted from your warm interior to the cold outside through both the glass and the solid frame. Double glazing slows this down by adding an extra pane of glass and an air pocket between the two panes. The ideal depth for this air pocket is 16mm. Triple glazing does an even better job. The speed of heat conduction through the frame is slowest for wooden frames.

The air that cools and falls at the window pane is a convection current. In double glazed units these currents exist in the air pocket in the middle as well as on the inside and outside of the glass. Filling the air pocket with a heavier gas such as argon reduces this convection and the heat loss that goes with it.

Just as light is radiated in to your room, so heat is radiated out. If, however, you add an invisible 'low emissivity' (low-e) coating to the window, much of this radiation will be reflected back into the room.

A poorly fitting window will leak air rapidly through all the gaps where the window meets the frame. Windows designed with good, long-lasting seals will keep the draughts out. When windows are replaced it is also important to seal the gap between the window frame and the wall with expanding foam or sealant.

The very best window

You will be doing well to beat a triple-glazed, argon-filled, low-e coated, well-sealed window. One detail worth looking out for is a 'warm edge'. Double and triple glazed windows have spacers along their edges which separate the panes. These are usually strips of aluminium which conduct heat rather too well and so create cold edges to the window panes. If you have condensation problems round the edges of your double-glazing, this is almost certainly why. When insulating spacers are used instead, this problem disappears.

The green choice for the window frame is wood. High quality timber windows made from sustainably sourced wood (page 132) are attractive, strong and durable. Softwoods will eventually require some maintenance but the first repaint may not be needed for a decade or more. If you want a hardwood frame, choose European oak rather than endangered tropical timber. Good quality, well-maintained timber frames will last much longer than plastic (uPVC) frames, which are toxic to manufacture and to dispose of.

Replace or renovate?

You will only get the very best performance from your windows if you have the very best windows fitted. However, replacing windows is expensive and may not be necessary if your current windows are in reasonable condition. You can still reduce the heat loss through your windows by draught-proofing (page 27) or, for a bigger outlay, getting the windows professionally renovated. The latter approach is recommended for old sash windows as these are difficult to draught-proof effectively if they do not fit properly.

Installing secondary glazing is another means of significantly reducing heat losses without sacrificing your existing windows. This is usually less expensive than replacing windows with new double glazing. Double-glazed argon-filled low-e glazing units in secondary wood frames will improve the performance of single-glazed windows by 300% and the end result will be even better than new double glazing. If you are on a budget you can improvise secondary glazing with removable sheets of transparent plastic.

Costs

The higher the window specification, the greater the cost. If however you need to replace your windows anyway, the extra cost of high performance windows is well worth considering, if only for the extra comfort that they will provide.

There are many UK suppliers of high quality timber windows. The Ecoplus range from the Green Building Store are made from FSC-certified timber and start at under £400 for opening casement windows (www.greenbuildingstore.co.uk, 01484 461705).

Further information

Talk to an energy efficiency advisor before you install new windows. Call the **Energy Saving Trust**'s national helpline on 0800 512 012 to be put in touch with a local advisor. The Trust's website also offers useful information on windows (www.energysavingtrust.org.uk) and its free online library of best practice guides includes *Windows for new and existing housing*.

The blue 'Energy Saving Recommended' label is now awarded to the best windows, so the Energy Saving Trust's online database of products is a good place to start looking for suppliers.

For detailed analysis of the performance of different window frame materials, see the **GreenSpec** website, www.greenspec.co.uk.

Wood

If you had to invent the perfect ecological building material, you would be hard pressed to come up with a better product than wood. This strong, light, flexible, durable material is made from atmospheric carbon dioxide using solar energy. It can be used in every part of the home, from floorboards to rafters. At the end of its life, it can return to the earth harmlessly. So use it with enthusiasm – but buy it with care.

Sourcing timber

Wood is only a good environmental choice if you are sure that it comes from a legal and sustainable source. If you nip down to your local timber yard to buy some plywood to board out your loft, you may be supporting the illegal destruction of the rainforest in south-east Asia. Plywood imported into Britain has been traced to sawmills in China which use hardwood veneers from ancient forests in Papua New Guinea. Across the globe, virgin forests are being destroyed to meet global demand for timber.

Timber 'chain of custody' certification makes it possible to buy wood products with confidence in their legal and sustainable source. The Forest Stewardship Council (FSC, www.fsc-uk.org) runs the most stringent and widely recognised certification scheme. Wood which comes with the FSC logo, or an FSC chain of custody (COC) number on the receipt, has been independently certified as coming from a well-managed forest.

In Europe there are many national timber certification schemes which come under the umbrella of the PEFC (Programme for the Endorsement of Forest Certification, www.pefc.co.uk). Timber from British forests is certified by both the FSC and PEFC against the UK Woodland Assurance Standard (www.ukwas.org.uk).

The FSC logo is now a familiar sight in big DIY stores. Local timber yards are catching up, so questions about chain of custody certification will not be unexpected. Do not accept a general assurance that all the timber on sale is 'eco-friendly'. If there is no logo, or no chain of custody documentation, you cannot be sure of this.

Reclaimed timber

There is a thriving market in reclaimed timber in Britain, especially tropical hardwoods such as teak that were once used in great quantities but are now scarce (though some new teak products can be bought with FSC certification). As reclaimed timber is the greenest of all wood choices, always consider whether you could use second-hand timber before you rush out to buy new.

Timber preservatives

There is also a thriving market in timber preservation in Britain but this is far from environmentally friendly. Look at the back of most tins of wood preservative or wood stain (which usually contains preservative) and you will see some ugly orange labels warning you of the toxicity of the product.

If wood is prepared and installed properly, preservatives are rarely necessary. Wet rot and dry rot will only take hold on moist wood, so if you use well-seasoned wood in well-ventilated spaces with no sustained exposure to damp, you are unlikely to suffer. Do everything you can to improve the ventilation of an affected area before you slap on preservatives. Some wood-boring insects can be more of a challenge to deal with, but even these will not survive in a warm, dry, centrally heated space. Avoid timber that includes sapwood, the living outer layers inside the tree, as this is vulnerable to insect attack.

When you do need to use a timber preservative, choose boron-based products as these have minimal health and environmental risks. If you are using timber outside, use species that have strong natural preservatives within them such as cedar, oak and Douglas fir.

If you are concerned about rot in the timber fabric of your house, or you have found evidence of insect attack, don't panic. Get a survey done by an independent surveyor and work out a long-term solution which will remove the conditions that created the problem. Bear in mind that holes in timbers may indicate an old, rather than current, problem with insect infestation.

Wood finishes

Many oils and varnishes also have nasty orange labels on the back. These can be avoided altogether by using products made from non-toxic, natural ingredients. These are usually based on linseed oil, a long established wood finish (page 63).

Costs

FSC and PEFC timber and wood products are now widely available and you should not have to pay more for them (though some suppliers still charge a premium).

Boron-based timber preservatives are available for both indoor and outdoor use. Five litres of borax interior wood preservative will cost you around £20 from the Green Building Store (www.greenbuildingstore.co.uk).

Further information

The UK's **Forest Stewardship Council** website has a directory of wood products and suppliers with FSC certification (www.fsc-uk.org). The organisation is happy to advise the general public on sourcing FSC products (01686 413916, info@fsc-uk.org). The international database of FSC suppliers is at www.fsc-info.org.

Information about the international trade in illegally logged timber can be found at www.illegal-logging.info.

Guides to treating rot and identifying wood-boring insect infestations are available on the **Safeguard** website (www.safeguardeurope.com).

Volume 2 of *The Green Building Handbook* (Tom Woolley and Sam Kimmins, E & F N Spon) describes both the content of wood preservatives and alternative strategies for protecting wood.

For up-to-date information about timber finishes and preservatives, see the **GreenSpec** website (www.greenspec.co.uk).

Wood burners

There is no better fuel to burn in your home than wood. When you burn wood, carbon dioxide is released into the atmosphere. But this is the same carbon dioxide that the tree absorbed when it was growing so there is no net change in the amount of greenhouse gas in the air. In contrast, when you burn fossil fuels that have been dug out of the ground, you are pushing the planet back to the over-heated state it was in millions of years ago. So burn wood and stay cool.

Finding fuel

Domestic wood-burning is possible anywhere in the country as some wood burners are permitted even in smokeless zones. The viability of using wood is primarily a matter of access to fuel. Although you can get logs delivered anywhere, there is little point in burning logs if a tank of diesel is needed to get them to you.

Many modern wood-burners use wood pellets instead of logs or woodchip. These are made from sawdust and other wood waste from industry. They are dense, dry and consistent in size, so they are a very effective fuel. The supply of wood pellets has been slow to pick up in Britain, resulting in an unhappy reliance on imports, but this situation is improving.

Whatever fuel you use, you need somewhere to store it. Logs must be left to dry for a couple of years before they are burnt, so you will need a proper woodshed if you want to season the wood yourself. Wood pellets come ready for use either in tidy bags or straight off the lorry into a hopper. The amount of storage space you require will depend on how much heat you are after but you will need a large hopper if wood pellets are to be your principal source of heat.

If you buy logs, you should make sure they come from a forest that is being constantly regenerated. Forest thinnings and coppiced timber are the best options as they are by-products of good forest management. Wood burning then becomes a means of providing economic support for the care and protection of woodland.

Burning choices

People use wood as a fuel for many different reasons and in many different ways. Do you want an open fire to stare into, or a warm centre to your home to supplement your central heating, or even all your heating and hot water to be fuelled by wood? The more you want from wood, the more important it is to consider the different options available and their implications for your daily life.

Some people love chopping wood and keep fit doing so. Most of us are less keen, especially on cold winter mornings. This can be avoided if you buy wood products that are ready to burn, but they still require a certain amount of lugging around. Lighting fires, sustaining them, moderating their heat and cleaning up ash afterwards can also be a tiresome and messy business. There are hi-tech ways of avoiding almost all these chores, but they will cost you. So in considering the specification for your heating system, reflect not only on its performance but also on how well your enthusiasm for wood-burning will stand up over the years – and how much automation you ought to invest in.

Wood-burning stoves

Wood-burning stoves are a popular way of burning all wood fuels. Many are highly efficient, and will retain and radiate heat long after the fire has died. A stove is a far more efficient way of burning wood than an open fire.

Some wood-burning stoves have back boilers which will provide hot water and even supply radiators. However a stove that goes out at night and leaves you with cold radiators in the morning is not ideal, so these systems often supplement standard oil or gas-fired central heating. Some systems supply 'thermal stores': special hot water cylinders that can keep the temperature up in the central heating when the fire has gone.

Automated wood pellet stoves are the most advanced wood-burning stoves. They have electronic ignition and thermostatic controls that automatically release fuel into the burner at just the rate to sustain the required temperature. They can supply hot water and central heating without the risk of going out when the heat is needed most. The wood pellet hopper at the top of the stove has to be replenished every 2 to 3 days.

Automated wood-pellet boilers

Automated wood pellet boilers are the final step in sophistication. These are not room heaters but are typically sited in utility rooms where they function like any boiler, providing heat to the radiators or hot water cylinder as the thermostatic controls and timer dictate. They are the most thoroughly automated technology, being fed from an external hopper that is directly replenished by the occasional truck. Log-burning boilers are also available but do not have the same level of automation as wood pellet boilers.

Costs

Wood is not an expensive fuel. Logs are cheaper than all the fossil fuel alternatives. Wood pellets are more expensive but comparable to gas. If you buy wood pellets by the tonne rather than by the bag you will get a better deal.

The simplest wood-burning stoves start at around £400. Automated wood-pellet stoves that supply hot water and central heating start at around £4,000. A fully installed automated wood-pellet boiler can cost from £7,000 to £12,000 or more.

See www.stovesonline.co.uk for prices of wood-burning stoves.

Grants are available to support the installation of automated wood pellet stoves and wood boilers. Contact the Energy Saving Trust to find out what you are eligible for (www.energysavingtrust.org.uk, 0800 512 012). If you apply for a grant, you have to use an installer registered with the scheme.

Further information

HETAS is the trade association for solid fuel burners and is a good place to look for qualified installers (www.hetas.co.uk).

The **Solid Fuel Association** offers advice and contacts (www.solidfuel.co.uk, 0845 601 4406).

For information on smoke control areas and the appliances that you can use within them, see www.uksmokecontrolareas.co.uk.

The Logpile Project maintains a database of wood fuel suppliers (www.nef.org.uk/logpile).

A free information sheet on *The Wood Fuelled Home* is available from the **Centre for Alternative Technology** (www.cat.org.uk, 01654 705989). For more detailed information, see *Home Heating with Wood* (Chris Laughton, CAT Publications).

Resources

AECB. The Sustainable Building Association (www.aecb.net). Invaluable network with excellent annual conference, online forum and online database of green professionals.

All Things Eco (www.allthingseco.co.uk). Directory of online eco retailers.

Centre for Alternative Technology (www.cat.org.uk). A long-established source of information, publications and advice. *Clean Slate* is CAT's regular magazine.

Energy Saving Trust (www.energysavingtrust.org.uk, 0800 512 012). An essential resource for everything to do with energy in the home, including up-to-date information on grants. The EST's free online library of best practice guides is excellent and extensive. Some guides are quite technical but many offer detailed practical advice about improving the energy efficiency of existing homes. See www.energysavingtrust.org.uk/housingbuildings/publications.

Encraft (www.encraft.co.uk). Offers home energy assessment and tailored advice on improving energy performance.

Environment Agency (www.environment-agency.gov.uk, 08708 506 506). A good source of information on water use, including the booklet *Conserving Water in Buildings*.

Forest Stewardship Council (www.fsc-uk.org). Website includes an online directory of sustainable timber suppliers.

Green Building Press (www.greenbuildingpress.co.uk). Publishes the *Green Building Magazine* and *Green Building Bible*. Hosts an online green product directory and the Green Building Forum.

Green Guides (www.greenguide.co.uk). Regional guides to green shopping and lifestyles.

GreenSpec (www.greenspec.co.uk). Invaluable guide to choosing and using greener materials.

Green Register (www.greenregister.org). Online register of green building professionals.

Low Impact Living Initiative (www.lowimpact.org). Information and courses on a wide range of eco-living topics.

National Energy Foundation (www.nef.org.uk). Online advice on energy efficiency and renewable energy technology.

National Home Energy Rating (www.nher.co.uk). Offers home energy assessment and tailored advice on improving energy performance.

The Nottingham Ecohome (www.msarch.co.uk/ecohome). An inspiring eco-renovation of a semi-detached Victorian house.

The Renewable Energy Centre (www.therenewableenergycentre.co.uk). Web portal of companies offering renewable energy and energy efficiency products.

The Yellow House (www.theyellowhouse.org.uk). A wonderful guide to a complete eco-renovation of a 1930s terraced house.

Top Ten (www.topten.ch). This guide to the most energy-efficient products on the market is the best place to start looking for new electrical and electronic goods. Not in English but easy to navigate.

Sourcing materials

Ecomerchant (www.ecomerchant.co.uk, 01795 530130).

Green Building Store (www.greenbuildingstore.co.uk, 01484 461705).

The Green Shop (www.greenshop.co.uk, 01452 770629).

Natural Building Technologies (www.natural-building.co.uk, 01844 338338).

Old House Store (www.oldhousestore.co.uk, 0118 969 7711).

Index

The main subject headings in this book are arranged alphabetically and listed in the Contents, and are not all replicated in the index.

Also available:

Diary of an Eco-Builder
by Will Anderson

"An entertaining read, and full of information too." – Architects Journal

Invaluable for anyone with an interest in eco-building or eco-renovating their home.
ISBN 978 1 903998 79 3 £14.95 paperback

The Green Self-Build Book
by Jon Broome

Inspiration and information to guide you through the green self-build process.
ISBN 978 1 903998 73 1 £25.00 paperback

In the same series:

For our complete book list, see www.greenbooks.co.uk